METHODS AND APPLICATIONS OF
WHITE NOISE ANALYSIS
IN INTERDISCIPLINARY SCIENCES

METHODS AND APPLICATIONS OF
WHITE NOISE ANALYSIS
IN INTERDISCIPLINARY SCIENCES

Christopher C Bernido
M Victoria Carpio-Bernido

Central Visayan Institute Foundation, Philippines

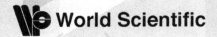

NEW JERSEY · LONDON · SINGAPORE · BEIJING · SHANGHAI · HONG KONG · TAIPEI · CHENNAI

Published by

World Scientific Publishing Co. Pte. Ltd.
5 Toh Tuck Link, Singapore 596224
USA office: 27 Warren Street, Suite 401-402, Hackensack, NJ 07601
UK office: 57 Shelton Street, Covent Garden, London WC2H 9HE

Library of Congress Cataloging-in-Publication Data
Bernido, Christopher C. (Christopher Caseñas)
  Methods and applications of white noise analysis in interdisciplinary sciences / by Christopher C. Bernido (Central Visayan Institute Foundation, Philippines), M. Victoria Carpio-Bernido (Central Visayan Institute Foundation, Philippines).
    pages cm
  Includes bibliographical references and index.
  ISBN 978-981-4569-11-8
  1. White noise theory. I. Carpio-Bernido, M. Victoria. II. Title.
  QA274.29B47 2015
  519.2'2--dc23

                    2014028642

British Library Cataloguing-in-Publication Data
A catalogue record for this book is available from the British Library.

Printed in Singapore

Contents

Preface

This book focuses on the infinite dimensional white noise analysis or Hida calculus invented by Takeyuki Hida in 1975. The presentation is from a practical viewpoint with the aim to augment the toolbox of physicists, chemists, and engineers, among others, for treating classical and quantum stochastic problems in interdisciplinary sciences. Chapter 2 introduces the basic concepts and methods of white noise analysis. Chapter 3 presents white noise analysis as a powerful tool in investigating systems with memory, especially when combined with the Feynman path integral method or summation-over-all-histories. Applications of the white noise calculus to non-Markovian physical and social systems begin in Chapter 4 with the evaluation of complex systems. This is followed by the analysis of a time series as modified Brownian motion with memory in Chapter 5. For Markovian fluctuations, the white noise approach is presented in Chapter 6. Such classes of applications are mostly discussed for the rest of the book. Applications in biophysics are presented in two chapters: single and multiple neuronal dynamics in Chapter 7 and biopolymer conformations in Chapter 8. Chapter 9 shifts the discussion to nonrelativistic quantum systems with a mathematically rigorous treatment of Feynman integrals. Chapter 10 deals with quantum problems with boundary conditions. Lastly, Chapter 11 illustrates some applications of white noise analysis in relativistic quantum mechanics.

We wish to thank Ludwig Streit and Takeyuki Hida for introducing us to white noise analysis. L. Streit also initiated us to fractional Brownian motion, during one of our visits at the Research Center BiBoS, Universität Bielefeld. Discussions with him were also carried out at the Centro de Ciencias Matematicas, Universidade da Madeira, and during his numerous visits to Jagna. Also acknowledged with thanks are enlightening discussions with

Tobias Kuna, José Luís da Silva, Maria João Oliveira, and Martin Grothaus on the mathematical aspects of white noise. We also remain grateful to Akira Inomata for our early introduction to path integrals as his thesis advisees at the State University of New York at Albany, to Hiroshi Ezawa for discussions during his visits to Jagna and our visits at the Gakushuin University in Tokyo, to Frederik W. Wiegel for his fruitful visits to Jagna, to Swee Cheng Lim for his valuable work on fractional dynamics and for being a staunch partner in the development of scientific research in our part of the world. These interactions in no small measure influenced our scientific outlook. We are also deeply grateful to the Alexander von Humboldt Foundation for research stays at the Research Center BiBoS and the Bernstein Center Freiburg, University of Freiburg, and to the Abdus Salam International Center for Theoretical Physics. C. C. B. acknowledges the support of the National Academy of Science and Technology (Philippines). Lastly, the authors' grateful appreciation goes to former students and present collaborators: Matthew G. O. Escobido, Jinky Bornales, Henry P. Aringa, Enrico Gravador, Nancy Lim, Cristine Villagonzalo, Glenn Aguarte and Robert Ramos, Jr., for the insightful discussions.

Chapter 1

Introduction

In recent years, rapid developments in advanced technologies have allowed the accumulation of large amounts of experimental and empirical data that could help provide a deeper understanding of many natural and social phenomena. Among these are experimental results coming from diverse areas: from neuroscience which probes problems ranging from tracking of memory and learning to Parkinson's disease, epilepsy, and schizophrenia; complex systems from biology to competitions in industry; and long-standing issues such as the protein folding problem. Clearly, the analysis, modelling, and simulation for better understanding of and predictive potential for these systems require powerful tools from different disciplines. Stochasticity, however, is an ubiquitous feature of real-world complex systems at different scales and dynamics. Often, non-Markovian properties and strong correlations compound already involved analytical computations for such systems. Nevertheless, more tractable approaches may be developed with mathematics beyond the common toolbox. We thus introduce in this book features of the white noise calculus introduced by Takeyuki Hida in 1975 as a novel approach to infinite dimensional analysis [103; 106]. The calculus, when combined with other methods such as Feynman's summation-over-all-histories [86], and fractional stochastic parametrization for non-Markovian processes, could open new avenues for resolving cross-disciplinary problems.

1.1 Some Properties of White Noise

Almost every system is subjected to oftentimes intractable external and internal influences which are referred to as noise or fluctuations. An early investigation of these fluctuations as a stochastic process was done in 1900 by Louis Bachelier who modelled stock market fluctuations in his Ph.D. thesis [12]. Although largely ignored in the beginning, this work of Bachelier has been considered as the birth of mathematical finance. This stochastic process was later referred to as Brownian motion $B(t)$, where t is time, as an offshoot of a work by Albert Einstein in 1905 [78; 204; 208]. Apparently unaware of the work of Bachelier, Einstein provided theoretical explanation for the fluctuating motion of a small particle immersed in fluid. This jittery motion of a particle in fluid was first observed by the botanist Robert Brown in 1827. Molecules of the fluid constantly collide with the particle in an unpredictable way so that the position of the particle fluctuates. A firm mathematical foundation for Brownian motion, however, was done only in 1923 by Norbert Wiener. His work, called the Wiener process, is often simply referred to as Brownian motion [90].

Although the exact position of a Brownian particle is difficult to pinpoint, it is possible to obtain a probability density of finding a particle in a given region. This probability density may be used to obtain expectation values which mathematically characterize Brownian motion. For example, one can obtain for the random continuous function $B(t)$ the results that:

(1) The $B(t)$ has mean zero, or its expectation value is,

$$E(B(t)) \equiv 0. \tag{1.1}$$

(2) Given a Brownian motion at two different times, $B(t)$ and $B(s)$, the expectation value of their product, or covariance is,

$$E(B(t)B(s)) = s, \tag{1.2}$$

for $s < t$. Equation (1.2) provides a measure of the correlation between $B(t)$ and $B(s)$.

If we take the time derivative of $B(t)$, the concept of Brownian motion directly leads to another random process involving the white noise variable $\omega(t)$. The $\omega(t)$ may informally be written as, $\omega(t) = dB(t)/dt$ [103; 105]. Since stochastic trajectories are not differentiable, a rigorous mathematical treatment in handling $\omega(t)$ would be needed. We can, however, view $\omega(t)$ as "velocity" of a Brownian motion. Alternatively, we may express Brownian motion as [90; 106],

$$B(t) = \int_0^t \omega(s) \, ds. \tag{1.3}$$

Compared with $B(t)$, the white noise random variable $\omega(t)$ has the following expectation values [90; 197]:

(1) The $\omega(t)$ also has mean zero, or expectation value,

$$E(\omega(t)) \equiv 0. \tag{1.4}$$

(2) Its covariance is given by a delta function,

$$E(\omega(t)\omega(s)) = \delta(s-t), \tag{1.5}$$

which is zero everywhere except when $s = t$, i.e., it is completely uncorrelated for different times and has independent values at each point.

One historical highlight in handling the random variable $\omega(t)$ as a mathematical object was the introduction of white noise calculus in 1975 [103]. Among many other applications, this calculus was then applied to Richard Feynman's path integral method, or "sum-over-all-histories" approach which has permeated almost all areas of phyiscs [86]. In 1983, a paper by Ludwig Streit and Takeyuki Hida [196] provided mathematical rigor to the Feynman path integral [86; 99; 185] and led to a series of applications of white noise analysis to quantum mechanical problems. The Hida-Streit [196] procedure has since then been successfully applied to treat harmonic [196] and anharmonic oscillators [100], a particle in a uniform magnetic field [61], in a Morse potential [63; 132], and in constrained systems such as a circular topology [21; 140], the Aharonov-Bohm set-up [21], and infinite well [21]. The method also works for time-dependent potentials [100]. The calculus allows the generalization of concepts in finite dimensions to the infinite dimensional case, including differential operators and integral transforms such as Fourier and Fourier-Mehler transforms [106]. Aside from being mathematically well-founded, the white noise approach is a practical tool for physicists. There is a rich potential beyond quantum mechanical problems for expanded application to other natural and social phenomena.

1.2 Preview of Applications

To enhance real-world applications, we allow for non-Markovian evolution of a system by considering a quantity x parametrized as,

$$x(\tau) = x_0 + \int_0^\tau f(\tau - t)\, g(t)\, dt \tag{1.6}$$

where x_0 is the initial value. The $f(\tau - t)$ serves as a memory function that describes how the value of $x(\tau)$ is affected by its history as t ranges from 0 to τ. The explicit forms of $f(\tau - t)$ and $g(t)$ in Eq. (1.6) can be chosen depending on the natural process being modelled. Several forms will be discussed in this book. For example, choosing a memory function given by,

$$f(\tau - t) = \frac{(\tau - t)^{\alpha - 1}}{\Gamma(\alpha)}, \qquad (1.7)$$

where $\Gamma(\alpha)$ is a Gamma function, casts Eq. (1.6) into the form,

$$x(\tau) = x_0 + \frac{1}{\Gamma(\alpha)} \int_0^\tau (\tau - t)^{\alpha - 1} \, g(t) \, dt. \qquad (1.8)$$

The second term in Eq. (1.8) has the general form of a fractional integral of arbitrary order α which in fractional calculus is denoted by an operator ${}_0D_\tau^{-\alpha}$ defined by [153; 192],

$$_0D_\tau^{-\alpha} g(t) = \frac{1}{\Gamma(\alpha)} \int_0^\tau (\tau - t)^{\alpha - 1} \, g(t) \, dt. \qquad (1.9)$$

The choice, Eq. (1.7) for $f(\tau - t)$, thus allows one to write the path parametrization Eq. (1.6) as,

$$x(\tau) = x_0 + \, _0D_\tau^{-\alpha} g(t). \qquad (1.10)$$

To model fluctuations occurring in natural and social phenomena, we now choose $g(t)$ in Eq. (1.6) as,

$$g(t) = h(t)\,\omega(t), \qquad (1.11)$$

to obtain

$$x(\tau) = x_0 + \int_0^\tau f(\tau - t) \, h(t)\,\omega(t)\,dt. \qquad (1.12)$$

In Eq. (1.12), the white noise variable $\omega(t)$ describes random fluctuations while $h(t)$ is a deterministic factor. The variable $x(\tau)$ may then be used to model changes in growth rates, metabolic rate fluctuations, a relative membrane potential in neurons, the conformation of a polymer, or paths of a quantum particle, among other possibilities. Equation (1.12) allows the quantity $x(\tau)$ to have memory of the past with $f(\tau - t)$ modulating the white noise variable $\omega(t)$ as t ranges from 0 to τ.

Alternatively, we may also write Eq. (1.12) as,

$$x(\tau) = x_0 + \int_0^\tau f(\tau - t) \ h(t) \ dB(t), \qquad (1.13)$$

where $dB(t) = \omega(t) \ dt$, with $B(t)$ an ordinary Brownian motion. If we again choose as an example Eq. (1.7) with $h(t) = 1$ and $\alpha = H + \frac{1}{2}$, Eq. (1.13) becomes,

$$x(\tau) = x_0 + B^H, \qquad (1.14)$$

where B^H is fractional Brownian motion in the Riemann-Liouville representation given by [156; 188],

$$B^H = \frac{1}{\Gamma\left(H + \frac{1}{2}\right)} \int_0^\tau (\tau - t)^{H - \frac{1}{2}} \ dB(t), \qquad (1.15)$$

with H the Hurst exponent. For fractional Brownian motion, there exists a wide range of possible applications [123; 156].

If one deals with systems with no memory of the past one takes, $f(\tau - t) = h(t) = 1$, in Eq. (1.13) which leads to,

$$x(\tau) = x_0 + B(\tau)$$

$$= x_0 + \int_0^\tau \omega(t) \, dt. \qquad (1.16)$$

Alternatively, Eq. (1.16) can be obtained from Eqs. (1.14) and (1.15) by choosing $H = 1/2$ as the value of the Hurst exponent. In discussing quantum mechanical path integrals, for example, Eq. (1.16) is used where paths starting at some initial point x_0 propagate as Brownian fluctuations $B(\tau)$. As seen from Eq. (1.16), velocity is given by, $dx(\tau)/d\tau = \omega(\tau)$, and the action S for path integrals in quantum mechanics becomes a functional of white noise variables $\omega(\tau)$ [196].

Given the path parametrization Eq. (1.12), the contributions of all possible histories, or paths $x(\tau)$ between two endpoints $x(0) = x_0$ and $x(T) = x_T$, can then be evaluated to yield the corresponding conditional probability density function of the system under investigation, or a propagator in quantum mechanics. Once the probability density function or quantum propagator is obtained, the system investigated is said to be essentially solved since evaluation of other interesting quantities would then be greatly facilitated. Starting in Chapter 3, we illustrate white noise path integration as a useful tool and technique in investigating varied physical situations. In particular, we discuss applications to problems encountered in complex systems, neurophysics, biophysics, polymer physics, time series analysis, and quantum systems under constraints or boundary conditions. Maximizing the utility of the Gaussian white noise variable $\omega(t)$, however, would require us to know certain aspects of white noise calculus.

1.3   A Calculus based on White Noise

The calculus of white noise has been developed extensively [106; 133; 135; 163]. A key feature of Hida's theory is to take the time derivative of Brownian motion $B(t)$, i.e., $\omega(t) = dB/dt$, where $\omega(t)$ is the white noise random variable, and treat the collection $\{\omega(t);\ t \in \mathbf{R}\}$ as a continuum coordinate system. Table 1.1 illustrates the correspondence between the finite-dimensional and infinite-dimensional cases [134].

Table 1.1. Correspondence between finite and infinite dimensions.

FINITE DIMENSIONS $\Rightarrow$	INFINITE DIMENSIONS
Independent variable: $x_j \Rightarrow$	*Independent random variable*: $\omega(\tau)$
Coordinate system: $(x_1, ..., x_n) \Rightarrow$	*Coordinate system*: $\{\omega(t);\ t \in \mathbf{R}\}$
Function: $f(x_1, ..., x_n) \Rightarrow$	*Functional*: $\Phi(\omega(t);\ t \in \mathbf{R})$
Space: $\mathbf{R^n} \Rightarrow$	*Space of Hida distributions*: S^*
Lebesgue measure: $dx \Rightarrow$	*Gaussian measure*: $d\mu(\omega)$

We give a brief introduction to white noise analysis in the next chapter with emphasis on concepts and tools needed for calculations. Topics are introduced initially at a simple level, addressed to those not familiar with the field, and developed from a practical perspective for possible applications in treating problems arising in diverse disciplines. The conditional probability density function for paths given by Eq. (1.12) is then evaluated in Chapter 3. Applications in various areas of current interest are explored in succeeding chapters.

Chapter 2

White Noise Analysis: Some Basic Notions and Terminology

We first outline some essential definitions and features of white noise analysis that will be used in succeeding sections to investigate various systems. We begin by observing that a stochastic process such as Brownian motion is usually described by the random variable X satisfying the stochastic differential equation,

$$dX = a(t, X) \, dt + b(t, X) \, dB(t) \,, \qquad (2.1)$$

where $a(t, X)$ and $[b(t, X)]^2$ are the drift and diffusion coefficients, respectively, and $B(t)$ is the Wiener process. The familiar Langevin equation follows as,

$$\dot{X} = a(t, X) + b(t, X) \, \omega(t) \,, \qquad (2.2)$$

where $\dot{X} = dX/dt$ and

$$\omega(t) = dB(t) \, / \, dt \,, \qquad (2.3)$$

is called the Gaussian white noise which may be viewed as the "velocity" of a Brownian motion.

Alternatively, Wiener's Brownian motion $B(\tau)$ is represented by,

$$B(\tau) = \int_0^\tau \omega(t) \, dt = \langle \omega, \mathbf{1}_{[0,\tau)} \rangle \,, \qquad (2.4)$$

where we define,

$$\langle \omega, \xi \rangle \equiv \int_0^\tau \omega(t)\xi(t) \, dt \,. \qquad (2.5)$$

7

It is the collection of infinitely many independent random variables, $\{\omega(t);\ t \in \mathbf{R}\}$, that Hida [106] treated as the coordinate system of an infinite dimensional space. Processes can then be described in terms of a generalized white noise functional, $\Phi\left(\omega(t);\ t \in \mathbf{R}\right)$, instead of a functional of Brownian motion, $f\left(B(t);\ t \in \mathbf{R}\right)$. The calculus allows the generalization of concepts in finite dimensions to the infinite dimensional case, including differential operators and integral transforms such as Fourier and Fourier-Mehler transforms [103]. This makes it particularly suitable for the evaluation of the Feynman path integral.

Here we simply present tools of white noise analysis useful for our applications. For rigorous construction and proofs, we refer the reader to the references [103; 106; 133; 163].

White noise analysis works with the Gelfand triple, $S \subset L^2 \subset S^*$, linking the spaces of test functions S and the Hida distributions S^* through a Hilbert space of square integrable functions L^2. By the Minlos theorem, the Hida white noise space $(S^*, \mathcal{B}, \mu)$ with probability measure μ, and σ-algebra $\mathcal{B}$ generated on S, is defined by the characteristic functional [106],

$$C(\xi) = \int_{S^*} \exp\left(i\left\langle \omega, \xi \right\rangle\right) d\mu(\omega) = \exp\left(-\frac{1}{2}\int_0^\tau \xi^2 dt\right). \qquad (2.6)$$

In Eq. (2.6), $\xi \in S$ is a test function and the white noise Gaussian measure is given by,

$$d\mu(\omega) = N_\omega \ \exp\left(-\frac{1}{2}\int \omega(t)^2 \ dt\right) \ d^\infty\omega\,, \qquad (2.7)$$

with the exponential responsible for the Gaussian fall-off, and N_ω is a normalization factor.

2.1 *T*- and *S*-Transforms

A description of white noise functionals can also be obtained through their T- and S-transforms [103; 106], which will later be shown useful in evaluating Feynman integrals. For instance, the T-transform for a generalized white noise functional $\Phi(\omega)$ is defined by,

$$T\Phi(\xi) \equiv \int_{S^*} \exp\left(i\left\langle \omega, \xi \right\rangle\right) \Phi(\omega) \ d\mu(\omega)\,, \qquad (2.8)$$

akin to an infinite-dimensional Gauss-Fourier transform. On the other hand, an S-transform is related to the T-transform as,

$$S\Phi(\xi) = C(\xi)\, T\Phi(-i\xi) \tag{2.9}$$

and

$$T\Phi(\xi) = C(\xi)\, S\Phi(i\xi)\,, \tag{2.10}$$

where $C(\xi)$ is given by Eq. (2.6).

2.2 Simple Examples

Let us consider several examples to familiarize ourselves with the S- and T-transforms of a white noise functional $\Phi(\omega)$ useful in deriving quantities of physical interest.

Example 2.1. $\Phi = 1$.

From Eqs. (2.6) and (2.8), we simply have the T-transform of Φ given by,

$$T\Phi(\xi) = \exp\left(-\frac{1}{2}\int \xi^2 d\tau\right) = C(\xi)\,. \tag{2.11}$$

Example 2.2. $\Phi(\omega) = \exp\left(i\langle \omega, \eta\rangle\right)$.

It is easy to see that from Eqs. (2.6) and (2.8), we obtain the T-transform,

$$T\Phi(\xi) = \exp\left(-\frac{1}{2}\int (\xi+\eta)^2 d\tau\right) = C(\xi+\eta)\,. \tag{2.12}$$

Example 2.3. $\Phi(\omega) = \exp\left(-i\langle \omega, \eta\rangle \sqrt{2}\, y\right)$.

Starting from Eq. (2.8), we have,

$$T\Phi(\xi) = \int_{S^*} \exp\left(i\left\langle\omega,\xi\right\rangle\right)\exp\left(-i\left\langle\omega,\eta\right\rangle\sqrt{2}\,y\right)\,d\mu(\omega)$$

$$= \int_{S^*} \exp\left(i\left\langle\omega,\xi-\eta\sqrt{2}\,y\right\rangle\right)\,d\mu(\omega)$$

$$= C(\xi - \eta\sqrt{2}\,y). \tag{2.13}$$

This result may be expressed in terms of the Hermite polynomials $H_k(x)$ [102]. To see this, we use Eq. (2.6) and write,

$$T\Phi(\xi) = C(\xi - \eta\sqrt{2}\,y) = \exp\left[-\frac{1}{2}\int\left(\xi-\eta\sqrt{2}\,y\right)^2 d\tau\right]$$

$$= \exp\left(-\frac{1}{2}\int \xi^2 d\tau\right)\exp\left[\sqrt{2}\,y\int\xi\eta\,d\tau - y^2\int\eta^2 d\tau\right]$$

$$= C(\xi)\,\exp\left[\sqrt{2}\,y\,\langle\xi,\eta\rangle - y^2\right], \tag{2.14}$$

where we took the norm, $\int\eta^2 d\tau = 1$. The exponential in Eq. (2.14) has a form similar to the generating function of the Hermite polynomials $H_k(x)$, and we obtain,

$$T\Phi(\xi) = C(\xi)\sum_{k=0}^{\infty}\frac{y^k}{k!}\,H_k\left(\langle\xi,\eta\rangle/\sqrt{2}\right). \tag{2.15}$$

Example 2.4. $\Phi(\omega) = H_k\left(\langle\omega,\eta\rangle/\sqrt{2}\right)$, where $\int\eta^2 dt = 1$.

Let us first consider the generating function for Hermite polynomials, $H_k\left(\langle\omega,\eta\rangle/\sqrt{2}\right)$, given by [103],

$$\exp\left[\sqrt{2}y\langle\omega,\eta\rangle - y^2\right] = \sum_{k=0}^{\infty}\frac{y^k}{k!}H_k\left(\langle\omega,\eta\rangle/\sqrt{2}\right). \tag{2.16}$$

If we take the T-transform of Eq. (2.16), we have,

$$T\left(\exp\left[\sqrt{2}y\langle\omega,\eta\rangle - y^2\right]\right)(\xi) = T\left(\sum_{k=0}^{\infty}\frac{y^k}{k!}H_k\left(\langle\omega,\eta\rangle/\sqrt{2}\right)\right)(\xi)$$

$$= \sum_{k=0}^{\infty}\frac{y^k}{k!}T\left(H_k\left(\langle\omega,\eta\rangle/\sqrt{2}\right)\right)(\xi). \tag{2.17}$$

Following the procedure of Example 2.3 leading to Eq. (2.13), the left-hand side of Eq. (2.17) is,

$$
T\left(\exp\left[\sqrt{2}y\left\langle\omega,\eta\right\rangle - y^2\right]\right)(\xi) = \int \exp\left(i\left\langle\omega,\xi\right\rangle\right)
$$

$$
\times \exp\left(\sqrt{2}y\left\langle\omega,\eta\right\rangle - y^2\right)d\mu\left(\omega\right)
$$

$$
= e^{-y^2}\int \exp\left(i\left\langle\omega,\xi - i\sqrt{2}y\eta\right\rangle\right)d\mu\left(\omega\right)
$$

$$
= e^{-y^2}C\left(\xi - i\sqrt{2}y\eta\right), \tag{2.18}
$$

where we used Eq. (2.6). This could be rewritten as,

$$
T\left(\exp\left[\sqrt{2}y\left\langle\omega,\eta\right\rangle - y^2\right]\right)(\xi) = \exp\left(-y^2\right)
$$

$$
\times \exp\left[-\frac{1}{2}\int_0^\tau\left(\xi - i\sqrt{2}y\eta\right)^2 dt\right]
$$

$$
= \exp\left(-y^2\right)\exp\left(-\frac{1}{2}\int_0^\tau \xi^2 dt\right)
$$

$$
\times \exp\left(i\sqrt{2}y\int_0^\tau \xi\,\eta\,dt\right)\exp\left(y^2\int_0^\tau \eta^2 dt\right)
$$

$$
= \exp\left(-\frac{1}{2}\int_0^\tau \xi^2 dt\right)\exp\left(i\sqrt{2}y\int_0^\tau \xi\,\eta\,dt\right)
$$

$$
= C\left(\xi\right)\exp\left(i\sqrt{2}y\int_0^\tau \xi\,\eta\,dt\right), \tag{2.19}
$$

where $\int \eta^2 dt = 1$. Using the result Eq. (2.19) for the left-hand side of Eq. (2.17), we obtain the equation,

$$
\sum_{k=0}^\infty \frac{y^k}{k!}T\left(H_k\left(\left\langle\omega,\eta\right\rangle/\sqrt{2}\right)\right)(\xi) = C\left(\xi\right)\exp\left(i\sqrt{2}y\int_0^\tau \xi\,\eta\,dt\right)
$$

$$
= C\left(\xi\right)\sum_{k=0}^\infty \frac{y^k}{k!}\left(i\sqrt{2}\int_0^\tau \xi\,\eta\,dt\right)^k. \tag{2.20}
$$

We see from Eq. (2.20) that the T-transform of Hermite polynomials is given by [103],

$$T\left(H_k\left(\langle\omega,\eta\rangle/\sqrt{2}\right)\right)(\xi) = C(\xi)\left(\left(i\sqrt{2}\right)\int_0^\tau \xi\,\eta\,dt\right)^k. \qquad (2.21)$$

2.3　The Gauss Kernel

An extremely useful example of a white noise functional is the Gauss kernel, $\Phi(\omega) = \mathcal{N}\exp\left(-\frac{1}{2}\langle\omega,k\omega\rangle\right)$. It is instructive to derive its S- and T-transform. Let us first take the S-transform of $\Phi(\omega)$ using Eqs. (2.6), (2.8) and (2.9), i.e.,

$$S\Phi(\xi) = \exp\left(-\frac{1}{2}\int \xi^2 d\tau\right)\mathcal{N}\int \exp\left(\langle\omega,\xi\rangle\right)\exp\left(-\frac{1}{2}\langle\omega,k\omega\rangle\right)d\mu(\omega). \qquad (2.22)$$

Using the expression for $d\mu(\omega)$ given by Eq. (2.7), we have,

$$S\Phi(\xi) = \mathcal{N}\int \exp\left(-\frac{1}{2}\int[\xi^2 - 2\omega\xi + (k+1)\omega^2]d\tau\right)N_\omega\ d^\infty\omega$$

$$= \mathcal{N}\int \exp\left[-\frac{k+1}{2}\int\left(\frac{\xi}{k+1}-\omega\right)^2 d\tau\right]$$

$$\times \exp\left(-\frac{k}{2(k+1)}\int\xi^2 d\tau\right)N_\omega\ d^\infty\omega. \qquad (2.23)$$

Shifting, $\omega \to \omega + [\xi/(k+1)]$, we get,

$$S\Phi(\xi) = \mathcal{N}\int \exp\left(-\frac{k}{2(k+1)}\int\xi^2 d\tau\right)\exp\left[-\frac{k+1}{2}\int\omega^2 d\tau\right]N_\omega\ d^\infty\omega$$

$$= \exp\left(-\frac{k}{2(k+1)}\int\xi^2 d\tau\right)\mathcal{N}\int \exp\left[-\frac{k}{2}\int\omega^2 d\tau\right]d\mu(\omega). \qquad (2.24)$$

From Eq. (2.24), we now take the S-transform of $\Phi(\omega) = \mathcal{N}\exp[-\frac{1}{2}\langle\omega,k\omega\rangle]$ as,

$$S\Phi(\xi) = \exp\left(-\frac{k}{2(k+1)}\int\xi^2 d\tau\right), \qquad (2.25)$$

by taking the normalization $\mathcal{N}$ to be,

$$\mathcal{N}^{-1} = \int \exp\left[-\frac{k}{2}\int \omega^2 d\tau\right] d\mu(\omega). \qquad (2.26)$$

Using Eqs. (2.10) and (2.25), the T-transform of $\Phi(\omega)$ can easily be shown to be,

$$T\Phi(\xi) = \exp\left[-\frac{1}{2}\langle\xi, (k+1)^{-1}\xi\rangle\right]. \qquad (2.27)$$

An important case relevant to the path integral is when $k = -(i+1)$, i.e., $\Phi(\omega) = I_0 = \mathcal{N} \exp[(\frac{i+1}{2})\int \omega(\tau)^2 d\tau]$ (see, e.g., Eq. (9.46)). From Eq. (2.27), its T-transform is,

$$TI_0(\xi) = \exp\left(-\frac{i}{2}\int \xi^2 d\tau\right). \qquad (2.28)$$

2.4 Donsker Delta Function

Another important example of a white noise functional Φ is a delta function whose argument contains the Brownian motion $B(t)$. One has, $\Phi = \delta(B(t) - a)$, which in the literature is referred to as a Donsker delta function. Let us consider its S- and T-transform. From Eq. (2.4), we may write, $\delta(B(t) - a) = \delta(\langle\omega, \mathbf{1}_{[0,t)}\rangle - a)$, $a \in \mathbf{R}$, and express the delta function in terms of its Fourier representation as [140],

$$\Phi = \delta\left(\langle\omega, \mathbf{1}_{[0,t)}\rangle - a\right) = \frac{1}{2\pi}\int \exp\left[i\lambda\left(\langle\omega, \mathbf{1}_{[0,t)}\rangle - a\right)\right] d\lambda. \qquad (2.29)$$

Taking the S-transform using Eqs. (2.6), (2.8) and (2.9) we have,

$$\begin{aligned}
S\Phi(\xi) &= \frac{1}{2\pi}\int\int C(\xi)\exp\left(i\langle\omega, -i\xi\rangle\right) \\
&\quad \times \exp\left[i\lambda\left(\langle\omega, \mathbf{1}_{[0,t)}\rangle - a\right)\right] d\lambda\, d\mu(\omega) \\
&= \frac{1}{2\pi}\exp\left(-\frac{1}{2}\int \xi^2 d\tau\right)\int\int \exp\left(i\langle\omega, -i\xi + \lambda\mathbf{1}_{[0,t)}\rangle\right) \\
&\quad \times \exp\left(-i\lambda a\right) d\lambda\, d\mu(\omega). \qquad (2.30)
\end{aligned}$$

Again referring to Eq. (2.6), we can write this as,

$$S\Phi(\xi) = \frac{1}{2\pi} \exp\left(-\frac{1}{2}\int \xi^2 d\tau\right) \int C\left(-i\xi + \lambda \mathbf{1}_{[0,t)}\right) \exp\left(-i\lambda a\right) \, d\lambda.$$

$$= \frac{1}{2\pi} \exp\left(-\frac{1}{2}\int \xi^2 d\tau\right) \int \exp\left[-\frac{1}{2}\int_0^t \left(-i\xi(\tau) + \lambda \mathbf{1}_{[0,t)}\right)^2 d\tau\right]$$
$$\times \exp\left(-i\lambda a\right) \, d\lambda$$

$$= \frac{1}{2\pi} \int \exp\left[-\frac{1}{2}\lambda^2 t + i\lambda \left(\int_0^t \xi(\tau)d\tau - a\right)\right] d\lambda. \qquad (2.31)$$

The Gaussian integral in Eq. (2.31) yields the result,

$$S\Phi(\xi) = \frac{1}{\sqrt{2\pi t}} \exp\left[-\frac{1}{2t}\left(\int_0^t \xi(\tau)d\tau - a\right)^2\right], \qquad (2.32)$$

for the Donsker delta function, $\Phi = \delta\left(B(t) - a\right)$. It is then a simple step to get from Eqs. (2.10) and (2.32), the corresponding T-transform given by,

$$T\Phi(\xi) = \frac{1}{\sqrt{2\pi t}} \exp\left[-\frac{1}{2}\int \xi^2 d\tau - \frac{1}{2t}\left(\int_0^t i\xi(\tau)d\tau - a\right)^2\right]. \qquad (2.33)$$

Note that the Donsker delta function is an example of a Hida distribution [106; 175].

2.5 Lévy's Stochastic Area

Let us next discuss Brownian motion on a plane and the area S_T it produces as investigated by P. Lévy [142]. In particular we evaluate, $\int \exp\left[izS_T\right] d\mu(\omega)$, by first considering a two-dimensional Brownian motion, $\mathbf{B}(t) = (B_x(t), B_y(t))$. Lévy's stochastic area S_T is defined as the area of the region enclosed by the Brownian curve produced by $\mathbf{B}(t)$, in the time interval $0 \le t \le T$, and the chord connecting the origin with the terminal point of $\mathbf{B}(t)$ (see Figure 2.1). This area is given by [104; 142],

$$S_T = \frac{1}{2}\int_0^T \left[B_x(t) \, dB_y(t) - B_y(t) \, dB_x(t)\right]. \qquad (2.34)$$

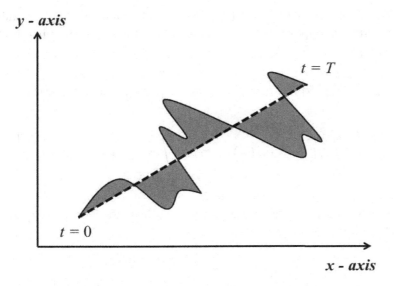

Fig. 2.1. For Brownian motion in two dimensions, Lévy's stochastic area (shaded part) is the region enclosed by a Brownian curve and the chord (dotted line) connecting the origin at $t = 0$, and the terminal point at $t = T$.

The two-dimensional Brownian motion $\mathbf{B}(t)$ can be realized on the probability space of a one-dimensional white noise by taking,

$$B_x(t) = \int_0^t \omega(\tau)\,d\tau \quad ; \quad dB_x(t) = \omega(t)\,dt \qquad (2.35)$$

$$B_y(t) = \int_{-t}^0 \omega(\tau)\,d\tau \quad ; \quad dB_y(t) = \omega(-t)\,dt. \qquad (2.36)$$

With Eqs. (2.35) and (2.36), we can write Eq. (2.34) as,

$$S_T = \frac{1}{2}\left[\int_0^T \left(\int_0^t \omega(\tau)\,d\tau\right)\omega(-t)\,dt - \int_0^T \left(\int_{-t}^0 \omega(\tau)\,d\tau\right)\omega(t)\,dt\right]$$

$$= \frac{1}{2}\left[\int_{-T}^0 \left(\int_0^{-t} \omega(\tau)\,d\tau\right)\omega(t)\,dt - \int_0^T \left(\int_{-t}^0 \omega(\tau)\,d\tau\right)\omega(t)\,dt\right].$$

$$(2.37)$$

We can now relabel t as τ_1, and τ as τ_2, and symmetrize S_T by adding terms with the opposite labels, i.e., t as τ_2, and τ as τ_1, and dividing the whole expression by two. We have,

$$
S_T = \frac{1}{4}\left[\int_{-T}^{0} d\tau_1 \int_{0}^{-\tau_1} d\tau_2\, w(\tau_2)\, w(\tau_1) + \int_{-T}^{0} d\tau_2 \int_{0}^{-\tau_2} d\tau_1\, w(\tau_1)\, w(\tau_2) \right.
$$
$$
\left. - \int_{0}^{T} d\tau_1 \int_{-\tau_1}^{0} d\tau_2\, w(\tau_2)\, w(\tau_1) - \int_{0}^{T} d\tau_2 \int_{-\tau_2}^{0} d\tau_1\, w(\tau_1)\, w(\tau_2) \right].
$$
(2.38)

If we now denote an integration over τ_1 with limits from 0 to T by the symbol $\chi_{[0,T]}(\tau_1)$, we can write (2.38) as,

$$
S_T = \int_{\mathbf{R}^2} w(\tau_1)\, F_S(\tau_1,\tau_2)\, w(\tau_2)\, d\tau_1 d\tau_2
$$
$$
= \langle w, F_S(\tau_1,\tau_2)\, w \rangle,
$$
(2.39)

where

$$
F_S(\tau_1,\tau_2) = \frac{1}{4}\left[\chi_{[-T,0]}(\tau_1)\, \chi_{[0,-\tau_1]}(\tau_2) + \chi_{[-T,0]}(\tau_2)\, \chi_{[0,-\tau_2]}(\tau_1) \right]
$$
$$
- \frac{1}{4}\left[\chi_{[0,T]}(\tau_1)\, \chi_{[-\tau_1,0]}(\tau_2) + \chi_{[0,T]}(\tau_2)\, \chi_{[-\tau_2,0]}(\tau_1) \right].
$$
(2.40)

The characteristic function of S_T (Theorem 4.9, p. 170 of [104]) is given by,

$$
\int \exp[izS_T]\, d\mu(w) = \left\{ \prod_{n=1}^{\infty}\left[1 + \frac{4z^2T^2}{4(2n-1)^2\pi^2} \right]^2 \right\}^{-1/2}
$$
$$
= [\cosh(zT/2)]^{-1}.
$$
(2.41)

With Eq. (2.39), we can write the exponential in Eq. (2.41) as, $\exp[izS_T] = \exp\left(-\frac{1}{2}\langle w, L\,w \rangle\right)$, where $L = -2izF_S(\tau_1,\tau_2)$. We can then write the left-hand side of Eq. (2.41) as,

$$
\int \exp\left(-\frac{1}{2}\langle w, L\,w \rangle\right) d\mu(w) = [\det(1+L)]^{-1/2},
$$
(2.42)

where we used Eq. (1.14) (with $\xi = 0$) of [196] to obtain the right-hand side. We also have from Eqs. (2.41) and (2.42) the relation,

$$[\det (1 + L)]^{-1/2} = [\cosh (zT/2)]^{-1}. \tag{2.43}$$

As discussed in Section 9.6, Lévy's stochastic area appears in the interaction term for a charged particle in a uniform magnetic field when the Feynman integrand is expressed as a white noise functional [61]. The evaluation of the integral, $\int \exp [izS_T]\, d\mu (\omega)$, is therefore helpful in handling the path integral for a charged particle in a uniform magnetic field, as well as other physical systems which exhibit the same mathematical form.

Exercises

(2-1) Show that the integral formula [103],

$$\int \exp (ia \langle \omega, \eta \rangle)\ H_k \left(\langle \omega, \eta \rangle / \sqrt{2} \right) d\mu (\omega) = \left(i\sqrt{2} \right)^k a^k \exp \left(-\frac{a^2}{2} \right),$$

is correct where $H_k \left(\langle \omega, \eta \rangle / \sqrt{2} \right)$ are Hermite polynomials and $\int \eta^2 dt = 1$.

(2-2) Consider the the Donsker delta function, $\Phi = \delta (B(t) - a)$, and its T-transform $T\Phi(\xi)$ given by Eq. (2.33).

(a) For $\xi = 0$, evaluate $\frac{\partial}{\partial t} [T\Phi(0)]$.

(b) For $\xi = 0$ and $a = x - x_0$, evaluate $\frac{\partial^2}{\partial x^2} [T\Phi(0)]$.

(c) From the results of (a) and (b) above, what type of equation is obeyed by $T\Phi(0)$?

(2-3) Show explicitly that Eq. (2.34) can be written as Eq. (2.39).

Chapter 3

Fluctuations with Memory

Many systems exhibit fluctuations arising from nonlinear interactions among multiple components with large numbers of degrees of freedom. The observed fluctuations of macroscopic quantities of these systems, however, could display regularities in behavior. These observed regularities compel us to look for general principles and universal features, in spite of not knowing detailed descriptions of multiple interactions among constituent components. For example, despite the seeming randomness of fluctuations, short-term and long-term memory could account for regularities in the dynamical evolution of macroscopic variables. In view of the huge number of physical and social systems characterized by such variables, it appears imperative to model these systems by incorporating memory in their dynamics.

3.1 Gibrat's Law

We begin this chapter by first discussing Gibrat's law which exemplifies fluctuations without memory. Consider a macroscopic quantity, labelled S, which fluctuates in time. To track the decrease or increase of S in time, we may start by looking at the simplest model of growth rates proposed by Gibrat [93]. One defines growth rate as, $R = S_1/S_0$, where S_0 and S_1 are values determined in two consecutive periods. Gibrat's law can then be used to relate growth rate at time t and a prior period $t - 1$ as,

$$R_t - R_{t-1} = \varepsilon_t R_{t-1} \,, \tag{3.1}$$

where ε_t is a random variable denoting the proportionate growth between the two periods. The uncorrelated random number ε_t has a mean close to zero. Equation (3.1) may be rewritten as,

$$
\begin{aligned}
R_t &= R_{t-1}\left(1 + \varepsilon_t\right) \\
&= R_{t-2}\left(1 + \varepsilon_{t-1}\right)\left(1 + \varepsilon_t\right) \\
&= R_0\left(1 + \varepsilon_1\right)\left(1 + \varepsilon_2\right)\ldots\left(1 + \varepsilon_t\right),
\end{aligned}
\tag{3.2}
$$

by expressing, $R_{t-1} = R_{t-2}\left(1 + \varepsilon_{t-1}\right)$, $R_{t-2} = R_{t-3}\left(1 + \varepsilon_{t-2}\right)$, and so on. Considering short-time periods where ε_i ($i = 1, 2, \ldots, t$) is small, we take the natural logarithm of Eq. (3.2) to get,

$$
\ln\left(R_t\right) \simeq \ln R_0 + \varepsilon_1 + \varepsilon_2 + \ldots + \varepsilon_t
\tag{3.3}
$$

where $\ln\left(1 + \varepsilon_i\right) \simeq \varepsilon_i$. If we define, $r = \ln R = \ln\left(S_1/S_0\right)$, we get [199],

$$
r_t \simeq r_0 + \sum_{i=1}^{t} \varepsilon_i,
\tag{3.4}
$$

where displacements $\varepsilon_1, \varepsilon_2, \varepsilon_3, \ldots$, are assumed independent and identically distributed random variables. Equation (3.4) then shows r_t to be a simple random walk starting at r_0 with t steps. One may also interpret this equation in analogy to a simple random walk on a periodic lattice where the length of each step is ε. At each step, the next jump may proceed toward any of the nearest-neighbor sites.

In the limit where the number of steps increases and the step sizes approach zero, one can pass over to a situation analogous to Eq. (3.4) but in continuous time using a Brownian motion $B\left(t\right)$ which starts at r_0 such that,

$$
\begin{aligned}
r\left(t\right) &= r_0 + B\left(t\right) \\
&= r_0 + \int_{t_0}^{t} \omega\left(s\right) ds,
\end{aligned}
\tag{3.5}
$$

with the expectation value, $\langle B\left(t\right)\rangle = 0$. In Eq. (3.5), $\omega\left(s\right)$ is a white noise variable or the derivative of the Brownian motion, i.e., $\omega\left(s\right) = dB/ds$.

Equations (3.4) and (3.5), however, are Markovian [75; 211]. This constrains applicability to real-world problems where short or long-memory processes may prevail in the evolution of complex systems. In the next section, we modify Eq. (3.5) by endowing growth rates of S with memory using a large class of parametrization larger than $B(t)$.

3.2 Parametrizing the Effects of Memory

To model a fluctuating variable with memory, we parametrize the evolution of a fluctuating quantity x by (see, e.g., Eq. (1.13)),

$$x(\tau) = x_0 + \int_0^\tau f(\tau - t) \ h(t) \ dB(t)$$

$$= x_0 + \int_0^\tau f(\tau - t) \ h(t) \ \omega(t)\, dt\,, \qquad (3.6)$$

where we used Eq. (2.3) and x_0 is the initial value. In Eq. (3.6), $B(t)$ is ordinary Brownian motion modified by a function $f(\tau - t) \ h(t)$. Unlike Eq. (3.5), the presence of $f(\tau - t)$ allows the quantity $x(\tau)$ to have memory of the past. As time t ranges from 0 to τ, the $f(\tau - t)$ modulates the evolution of the white noise variable $\omega(t)$ thereby affecting the value or history of $x(\tau)$. The $f(\tau - t)$, therefore, serves as a memory function and its explicit form, as well as that of $h(t)$, can be chosen depending on the process being modeled.

The probability density function for fluctuations with memory can also be evaluated. Let us consider the variable $x(\tau)$ whose evolution is described by Eq. (3.6). One could have an ensemble of all possible paths which start at x_0 at time $\tau = 0$ and ask what the probability would be that these paths end at a specific endpoint, $x(T) = x_T$, at a later time $\tau = T$. Following Feynman's sum-over-all possible histories [46; 86], we consider all paths $x(\tau)$ which satisfy the δ-function constraint of the form, $\delta(x(T) - x_T)$, where $x(T)$ is given by Eq. (3.6) for $\tau = T$, i.e.,

$$x(T) = x_0 + \int_0^T f(T - t) \ h(t)\, \omega(t)\, dt\,. \qquad (3.7)$$

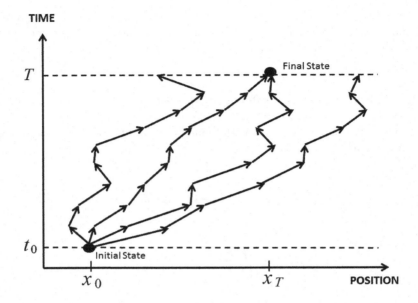

Fig. 3.1. At time $t = T$, paths may end at different points in space. The delta function constraint for paths on the left and right gives zero. The two paths in the middle contribute to the probability density function.

The probability density function $P(x_T, T; x_0, 0)$ for paths satisfying the δ-function constraint can be obtained by evaluating the expectation value $E\left(\delta\left(x\left(T\right) - x_T\right)\right)$, i.e.,

$$P\left(x_T, T; x_0, 0\right) = E\left(\delta\left(x\left(T\right) - x_T\right)\right)$$

$$= \int \delta\left(x\left(T\right) - x_T\right)\ d\mu$$

$$= \int \delta\left(x_0 + \int_0^T f\left(T - t\right)\ h\left(t\right)\ \omega\left(t\right) dt - x_T\right) d\mu,$$

$$(3.8)$$

where we used Eq. (3.6) and $d\mu$ is the Gaussian white noise measure [106] (see Figure 3.1).

Writing the delta function in terms of its Fourier representation we have,

$$P\left(x_T, T; x_0, 0\right) = \frac{1}{2\pi} \int d\mu \int\limits_{-\infty}^{+\infty} dk$$

$$\times \exp\left\{ik\left[\left(x_0 - x_T + \int\limits_0^T f\left(T-t\right)\, h\left(t\right)\, \omega\left(t\right) dt\right)\right]\right\}$$

$$= \frac{1}{2\pi} \int\limits_{-\infty}^{+\infty} dk \exp\left\{ik\left[\left(x_0 - x_T\right)\right]\right\}$$

$$\times \int \exp\left\{ik \int\limits_0^T f\left(T-t\right) h\left(t\right) \omega\left(t\right) dt\right\} d\mu. \qquad (3.9)$$

We can use Eq. (2.6) for the integration over $d\mu$, i.e.,

$$\int \exp\left\{i \int\limits_0^T \omega\left(t\right) \xi\left(t\right)\, dt\right\} d\mu = \exp\left\{-\frac{1}{2}\int\limits_0^T \xi^2\left(t\right)\, dt\right\}, \qquad (3.10)$$

where $\xi\left(t\right) = k\, f\left(T-t\right)\, h\left(t\right)$. Using this result in Eq. (3.9) we have,

$$P\left(x_T, T; x_0, 0\right) = \frac{1}{2\pi} \int\limits_{-\infty}^{+\infty} \exp\left\{ik\left[\left(x_0 - x_T\right)\right] - \frac{k^2}{2}\int\limits_0^T \left[f\left(T-t\right) h\left(t\right)\right]^2 dt\right\} dk.$$

$$(3.11)$$

The remaining integral is a Gaussian integral which can be evaluated to yield the form,

$$P\left(x_T, T; x_0, 0\right) = \left(2\pi \int\limits_0^T \left[f\left(T-t\right)\, h\left(t\right)\right]^2 dt\right)^{-\frac{1}{2}}$$

$$\times \exp\left(-\left[\int\limits_0^T \left[f\left(T-t\right)\, h\left(t\right)\right]^2 dt\right]^{-1} \frac{\left(x_T - x_0\right)^2}{2}\right).$$

$$(3.12)$$

Further simplification of the probability density function, Eq. (3.12), would depend on the explicit choice of the memory function $f(T-t)$ and $h(t)$. Take, for instance, the special case where $f(T-t)$ is just a constant, say, $f = \sqrt{2D}$ where D is a diffusion coefficient and $h(t) = 1$. Then, Eq. (3.12) reduces to the Gaussian distribution,

$$P(x_T, T; x_0, 0) = \frac{1}{\sqrt{4\pi DT}} \exp\left(\frac{-(x_T - x_0)^2}{4DT}\right), \qquad (3.13)$$

which solves the diffusion equation for the Wiener process.

In general, once the probability density function $P(x_T, T; x_0, 0)$ is solved, other interesting properties such as the mean square displacement, standard deviation, and probability of finding the state of a system in a particular domain can be evaluated, among others. This would be especially useful at present given the large volumes of data available online accumulated from astrophysical observations, genomics, telecommunications, urban transport development and management, and economic demographics, among others. Systematic analytics could entail pattern seeking, investigation of collective and individual behavior for multi-component systems, some or all of which may involve memory or strong correlations.

There are many possible choices for $f(T-t)\,h(t)$ which could allow a wide array of applications in many areas. In the next section, we consider a number of examples of $f(T-t)\,h(t)$ combinations for which closed form solutions can be obtained.

3.3   Memory Functions and Probability Densities

In modelling a fluctuating quantity $x(\tau)$, we have parametrized it as Eq. (3.6) where the $f(\tau - t)\,h(t)$ modifies the ordinary Brownian motion $B(t)$ at each step as the system progresses from an initial time 0 to a given time τ. The probability that $x(\tau) = x_T$, at time $\tau = T$, if the system started at x_0 would then be described by a probability density function $P(x_T, T; x_0, 0)$. We summarize in Table 3.1 some choices of $f(T-t)$. The equation numbers in the third column refer to references [3] and [98] needed to obtain the corresponding probability density function, Eq. (3.12).

In the following sections, we briefly discuss several memory functions found in Table 3.1.

Table 3.1. Memory function with corresponding Probability Density Function. Equation numbers are those in references [3] and [98].

Memory Function $f(T-t)$	$h(t)$	Probability Density Function $P(x_T, T; x_0, 0)$
(1) $f = \frac{(T-t)^{H-1/2}}{\Gamma(H+1/2)}$	$h = 1$	$\sqrt{\dfrac{H\,\Gamma^2\left(H+\frac{1}{2}\right)}{\pi T^{2H}}}$ $\times \exp\left(-\dfrac{H\,\Gamma^2\left(H+\frac{1}{2}\right)(x_0-x_T)^2}{T^{2H}}\right)$
(2) $f = \sin^{\frac{1}{2}}(T-t)$	$h = \sqrt{J_0(t)}$	$[2\pi T J_1(T)]^{-\frac{1}{2}} \exp\left(-\dfrac{(x_T-x_0)^2}{2TJ_1(T)}\right)$ Eq. (6.674.7) of [98]
(3) $f = \cos^{\frac{1}{2}}(T-t)$	$h = \sqrt{J_0(t)}$	$[2\pi T J_0(T)]^{-\frac{1}{2}} \exp\left(-\dfrac{(x_T-x_0)^2}{2TJ_0(T)}\right)$ Eq. (6.674.8) of [98]
(4) $f = (T-t)^{\frac{\mu-1}{2}}$ [Re $\mu > 0$, $T > 0$]	$h = \dfrac{e^{-\beta/2t}}{t^{(\mu+1)/2}}$	$\dfrac{\beta^{\mu/2}e^{\beta/2T}}{\sqrt{2\pi}\Gamma(\mu)T^{\mu-1}} \exp\left(-\dfrac{\beta^\mu e^{\beta/T}(x_T-x_0)^2}{2\,\Gamma(\mu)\,T^{\mu-1}}\right)$ Eq. (3.471.3) of [98]
(5) $f = (T-t)^{\frac{\mu-1}{2}}$ [Re $\mu > 0$, $T > 0$]	$h = \dfrac{e^{-\beta/2t}}{t^{(1-\nu)/2}}$ [Re $\beta > 0$]	$\dfrac{T^{\frac{1-2\mu-\nu}{4}}e^{\frac{\beta}{4T}}}{\sqrt{2\pi\beta^{\frac{\nu-1}{2}}\Gamma(\mu)W_{\frac{1-2\mu-\nu}{2},\frac{\nu}{2}}\left(\frac{\beta}{T}\right)}}$ $\times \exp\left(\dfrac{-\beta^{\frac{1-\nu}{2}}T^{\frac{1-2\mu-\nu}{2}}e^{\frac{\beta}{2T}}(x_T-x_0)^2}{2\Gamma(\mu)W_{\frac{1-2\mu-\nu}{2},\frac{\nu}{2}}\left(\frac{\beta}{T}\right)}\right)$ Eq. (3.471.2) of [98]
(6) $f = (T-t)^{\frac{\mu-1}{2}}$ [Re $\mu > 0$, $T > 0$]	$h = \dfrac{e^{-\beta/2t}}{t^\mu}$ [Re $\beta > 0$]	$\dfrac{1}{\sqrt{2\sqrt{\frac{\pi}{T}}\beta^{\frac{1}{2}-\mu}e^{-\frac{\beta}{2T}}\Gamma(\mu)K_{\mu-\frac{1}{2}}\left(\frac{\beta}{2T}\right)}}$ $\times \exp\left(\dfrac{-\sqrt{\pi T}(x_T-x_0)^2}{2\beta^{\frac{1}{2}-\mu}e^{-\frac{\beta}{2T}}\Gamma(\mu)K_{\mu-\frac{1}{2}}\left(\frac{\beta}{2T}\right)}\right)$ Eq. (3.471.4) of [98]
(7) $f = (T-t)^{\frac{\mu-1}{2}}$ [Re $\mu > 0$]	$h = \dfrac{e^{\beta t/2}}{t^{(1-\nu)/2}}$ [Re $\nu > 0$]	$\dfrac{1}{\sqrt{2\pi B(\mu,\nu)T^{T+\nu-1}{}_1F_1(\nu;\mu+\nu;\beta T)}}$ $\times \exp\left(\dfrac{-(x_T-x_0)^2}{2B(\mu,\nu)T^{T+\nu-1}{}_1F_1(\nu;\mu+\nu;\beta T)}\right)$ Eq. (3.383.1) of [98]
(8) $f = (T-t)^{\frac{\mu-1}{2}}$ [Re $\mu > 0$]	$h = \dfrac{e^{\beta t/2}}{t^{(1-\mu)/2}}$	$\dfrac{1}{\sqrt{2\pi^{\frac{3}{2}}\left(\frac{T}{\beta}\right)^{T-\frac{1}{2}}e^{\left(\frac{\beta T}{2}\right)}\Gamma(\mu)I_{\mu-\frac{1}{2}}\left(\frac{\beta T}{2}\right)}}$ $\times \exp\left(\dfrac{-\left(\frac{T}{\beta}\right)^{\frac{1}{2}-T}(x_T-x_0)^2}{2\sqrt{\pi}e^{\left(\frac{\beta T}{2}\right)}\Gamma(\mu)I_{\mu-\frac{1}{2}}\left(\frac{\beta T}{2}\right)}\right)$ Eq. (3.383.2) of [98]
[Re $\mu > 0$]		

Table 3.1. (Cont'd)

Memory Function $f(T-t)$	$h(t)$	Probability Density Function $P(x_T, T; x_0, 0)$
(9) $f = (T-t)^{\frac{\mu-1}{2}}$ [Re $\mu > 0$]	$h = \dfrac{\sin^{\frac{1}{2}}(at)}{t^{\frac{1-\mu}{2}}}$	$\sqrt{\dfrac{\pi^{-3/2}T^{\frac{1}{2}-\mu}a^{\mu-\frac{1}{2}}}{2\sin(\frac{aT}{2})\Gamma(\mu)J_{\mu-\frac{1}{2}}(\frac{aT}{2})}}$ $\times \exp\left(-\dfrac{\pi^{-\frac{1}{2}}T^{\frac{1}{2}-\mu}a^{\mu-\frac{1}{2}}(x_0-x_T)^2}{2\sin(\frac{aT}{2})\Gamma(\mu)J_{\mu-\frac{1}{2}}(\frac{aT}{2})}\right)$ Eq. (3.768.7) of [98]
(10) $f = (T-t)^{\frac{\mu-1}{2}}$ [Re $\mu > 0$]	$h = \dfrac{\cos^{\frac{1}{2}}(at)}{t^{\frac{1-\mu}{2}}}$	$\sqrt{\dfrac{\pi^{-3/2}T^{\frac{1}{2}-\mu}a^{\mu-\frac{1}{2}}}{2\cos(\frac{aT}{2})\Gamma(\mu)J_{\mu-\frac{1}{2}}(\frac{aT}{2})}}$ $\times \exp\left(-\dfrac{\pi^{-\frac{1}{2}}T^{\frac{1}{2}-\mu}a^{\mu-\frac{1}{2}}(x_0-x_T)^2}{2\cos(\frac{aT}{2})\Gamma(\mu)J_{\mu-\frac{1}{2}}(\frac{aT}{2})}\right)$ Eq. (3.768.9) of [98]
(11) $f = (T-t)^{\frac{\mu-1}{2}}$ [Re $\mu > 0$]	$h = \dfrac{(t^2+\beta^2)^{\frac{\nu}{2}}}{t^{\frac{1-\lambda}{2}}}$ [$\lambda > 0$]	$\sqrt{\dfrac{\beta^{-2\nu}T^{1-\lambda-\mu}}{2\pi B(\lambda,\mu)\,_3F_2\left(-\nu,\frac{\lambda}{2},\frac{\lambda+1}{2};\frac{\lambda+\mu}{2},\frac{\lambda+\mu+1}{2};\frac{-T^2}{\beta^2}\right)}}$ $\times \exp\left(\dfrac{-[B(\lambda,\mu)]^{-1}\beta^{-2\nu}T^{1-\lambda-\mu}(x_0-x_T)^2}{2\,_3F_2\left(-\nu,\frac{\lambda}{2},\frac{\lambda+1}{2};\frac{\lambda+\mu}{2},\frac{\lambda+\mu+1}{2};\frac{-T^2}{\beta^2}\right)}\right)$ $\left[\text{Re}\left(\frac{T}{\beta}\right) > 0\right]$; Eq. (3.254.1) of [98]
(12) $f = (T-t)^{-\frac{\nu}{2}}$ [$|\text{Re }\nu| < 1$]	$h = \sqrt{\dfrac{t^\nu}{(t-c)}}$ [$c < T$]	$\dfrac{1}{\pi\sqrt{2\csc(\nu\pi)\left[1-\cos(\nu\pi)\left(\frac{c}{T-c}\right)^\nu\right]}}$ $\times \exp\left(\dfrac{-\csc^{-1}(\nu\pi)(x_T-x_0)^2}{2\pi\left[1-\cos(\nu\pi)\left(\frac{c}{T-c}\right)^\nu\right]}\right)$ Eq. (3.228.1) of [98]
(13) $f = (T-t)^{-\nu/2}$ [$0.5 < \text{Re }\nu < 1$]	$h = \sqrt{\dfrac{t^{\nu-1}}{(t-c)}}$ [$c < T$]	$\sqrt{\dfrac{(T-c)^\nu}{-2\pi^2(c)^{\nu-1}\cot(\nu\pi)}}$ $\times \exp\left(\dfrac{(T-c)^\nu(x_T-x_0)^2}{2\pi(c)^{\nu-1}\cot(\nu\pi)}\right)$ Eq. (3.228.2) of [98]
(14) $f = (T-t)^{\nu/2}$ [Re $\nu > -1$, $T > 0$]	$h = e^{-\mu t/2}$	$\dfrac{1}{\sqrt{2\pi(-\mu)^{-\nu-1}e^{-T\mu}\,\gamma(\nu+1,-T\mu)}}$ $\times \exp\left(\dfrac{-e^{T\mu}(x_T-x_0)^2}{2(-\mu)^{-\nu-1}\gamma(\nu+1,-T\mu)}\right)$ Eq. (3.382.1) of [98]
(15) $f = \sqrt{J_{1-\nu}(T-t)}$ [$-1 < \text{Re }\nu < 2$]	$h = \sqrt{J_\nu(t)}$	$\dfrac{1}{\sqrt{2\pi(J_0(T)-\cos T)}}$ $\times \exp\left(\dfrac{-(x_T-x_0)^2}{2(J_0(T)-\cos T)}\right)$ Eq. (11.3.38) of [3]
(16) $f = \sqrt{J_\nu(T-t)}$ [Re $\nu > -1$]	$h = \sqrt{t^{-1}J_\mu(t)}$ [Re $\mu > 0$]	$\sqrt{\dfrac{\mu}{2\pi J_{\mu+\nu}(T)}}$ $\times \exp\left(\dfrac{-\mu(x_T-x_0)^2}{2J_{\mu+\nu}(T)}\right)$ Eq. (11.3.40) of [3]

3.4 Fractional Brownian Motion

A familiar case of a memory function is,

$$f(T - t) = \frac{(T - t)^{H-1/2}}{\Gamma(H + 1/2)}, \tag{3.14}$$

where, $h(t) = 1$, and H is the Hurst index ($0 < H < 1$) [110; 156]. With this choice of $f(T - t)$, the probability density function, Eq. (3.12), becomes [46]

$$P(x_T, T; x_0, 0) = \sqrt{\frac{H\,\Gamma^2\left(H + \frac{1}{2}\right)}{\pi T^{2H}}} \exp\left\{-\frac{H\,\Gamma^2\left(H + \frac{1}{2}\right)(x_0 - x_T)^2}{T^{2H}}\right\}. \tag{3.15}$$

Note that Eq. (3.14) allows us to write the parametrization of a fluctuating quantity, Eq. (3.6), as $x(\tau) = x_0 + B^H(\tau)$, where $B^H(\tau)$ is a fractional Brownian motion in the Riemann-Liouville fractional integral representation defined by [156],

$$B^H(\tau) = \frac{1}{\Gamma\left(H + \frac{1}{2}\right)} \int_0^\tau (\tau - t)^{H-1/2}\, dB(t). \tag{3.16}$$

Of the examples of a memory function, the one for fractional Brownian motion is the most mathematically developed and its application to natural processes has been widespread [37; 123; 124; 141; 145; 156; 188; 203].

3.5 Bessel-modified Brownian Motion

An oscillatory form for the memory function can also be given by,

$$f(T - t)\ h(t) = \sin^{\frac{1}{2}}(T - t)\ \sqrt{J_0(t)}, \tag{3.17}$$

where $J_\nu(t)$ is a Bessel function of the first kind. With Eq. (6.674.7) of [98], this yields from Eq. (3.12) the probability density function,

$$P(x_T, T; x_0, 0) = \frac{1}{\sqrt{2\pi T J_1(T)}} \exp\left(-\frac{(x_T - x_0)^2}{2T J_1(T)}\right). \tag{3.18}$$

Note that, for this case, Eq. (3.6) may be written as $x(\tau) = x_0 + B^J(\tau)$ where

$$B^J(\tau) = \int_0^\tau \sin^{\frac{1}{2}}(\tau - t) \sqrt{J_0(t)} dB(t).$$ (3.19)

Equation (3.19) shows that $B^J(\tau)$ results from the Bessel function and sine function modulating the derivative of ordinary Brownian motion, $dB(t)$, as t varies from 0 to τ.

3.6   Exponentially-modified Brownian Motion

Although there are many others, our last example is the combination,

$$f(T - t)h(t) = (T - t)^{(\mu-1)/2} \left(\frac{e^{-\beta/2t}}{t^{(\mu+1)/2}} \right).$$ (3.20)

Using Eq. (3.471.3) of [98], the probability density function Eq. (3.12) becomes,

$$P(x_T, T; x_0, 0) = \frac{\beta^{\mu/2} e^{\beta/2T}}{\sqrt{2\pi\Gamma(\mu) T^{\mu-1}}} \exp\left(-\frac{\beta^\mu e^{\beta/T} (x_T - x_0)^2}{2 \Gamma(\mu) T^{\mu-1}} \right),$$ (3.21)

where $\mu > 0$, and $\Gamma(\mu)$ is the Gamma function. For this case the path parametrization, Eq. (3.6), may be written as $x(\tau) = x_0 + B^{\mu,\beta}(\tau)$ where,

$$B^{\mu,\beta}(\tau) = \int_0^\tau \frac{(\tau - t)^{(\mu-1)/2}}{t^{(\mu+1)/2} e^{\beta/2t}} dB(t).$$ (3.22)

In Eq. (3.22), the factor in front of $dB(t)$ modifies the behavior of ordinary Brownian diffusion. The diffusion may be categorized into short-memory processes for $0 < \mu < 1$, and long-memory for $1 < \mu$.

3.7   Moments of the Probability Density Function

In Sections 3.3 to 3.6 we looked at several types of memory functions $f(T - t)$ that can be used together with the probability density function $P(x_T, T; x_0, 0)$ given by Eq. (3.12). We now examine the moments of

$P\left(x_T, T; x_0, 0\right)$ where the moment of nth order is defined as (denoting the endpoint simply as, $x_T = x$) [147],

$$\langle x^n \rangle = \int\limits_{-\infty}^{+\infty} x^n P\left(x, T; x_0, 0\right) dx\,. \qquad (3.23)$$

Consider the moment of zeroth order of the form,

$$\langle x^0 \rangle = \int\limits_{-\infty}^{+\infty} P\left(x, T; x_0, 0\right)\, dx$$

$$= \int\limits_{-\infty}^{+\infty} \left(2\pi \int\limits_0^T \left[f\left(T - t\right) h\left(t\right)\right]^2 dt \right)^{-\frac{1}{2}}$$

$$\times \exp\left\{ -\left(\int\limits_0^T \left[f\left(T - t\right) h\left(t\right)\right]^2 dt \right)^{-1} \frac{\left(x - x_0\right)^2}{2} \right\} dx\,, \qquad (3.24)$$

where we used Eq. (3.12). To facilitate evaluation we let,

$$y = x - x_0 \qquad ; \qquad dy = dx \qquad (3.25)$$

$$\alpha^2 = \frac{1}{2 \int\limits_0^T \left[f\left(T - t\right) h\left(t\right)\right]^2 dt} \qquad (3.26)$$

to obtain,

$$\langle x^0 \rangle = \frac{\alpha}{\sqrt{\pi}} \int\limits_{-\infty}^{+\infty} \exp\left(-\alpha^2 y^2\right)\, dy$$

$$= \frac{\alpha}{\sqrt{\pi}} \frac{\sqrt{\pi}}{\alpha} = 1\,. \qquad (3.27)$$

This result for the moment of zeroth order simply states the normalization,

$$\int\limits_{-\infty}^{+\infty} P\left(x, T; x_0, 0\right) dx = 1\,. \qquad (3.28)$$

With Eq. (3.28), it may also be good to recall that $P\left(x, T; x_0, 0\right)$ is the expectation value of a delta function as shown in Eq. (3.8).

Next, we look at the first moment of the probability density function which is just the mean or expectation value of x, i.e., $\langle x^1 \rangle = \langle x \rangle$, of the form,

$$
\langle x \rangle = \int_{-\infty}^{+\infty} x \, P(x, T; x_0, 0) \, dx
$$

$$
= \left(2\pi \int_0^T [f(T-t) \, h(t)]^2 \, dt \right)^{-\frac{1}{2}}
$$

$$
\times \int_{-\infty}^{+\infty} x \, \exp\left(-\left[\int_0^T [f(T-t) \, h(t)]^2 \, dt \right]^{-1} \frac{(x-x_0)^2}{2} \right) dx.
$$

$$(3.29)$$

Again, by using the substitutions, Eqs. (3.25) and (3.26), we get,

$$
\langle x \rangle = \frac{\alpha}{\sqrt{\pi}} \int_{-\infty}^{+\infty} (y + x_0) \exp\left(-\alpha^2 y^2 \right) \, dy
$$

$$
= \frac{\alpha}{\sqrt{\pi}} \int_{-\infty}^{+\infty} y \, \exp\left(-\alpha^2 y^2 \right) \, dy + \frac{\alpha x_0}{\sqrt{\pi}} \int_{-\infty}^{+\infty} \exp\left(-\alpha^2 y^2 \right) dy
$$

$$
= 0 + \frac{\alpha x_0}{\sqrt{\pi}} \frac{\sqrt{\pi}}{\alpha} = x_0,
$$

$$(3.30)$$

where the first integral is symmetric around y and yields zero. The mean $\langle x \rangle$, therefore, is just the initial point x_0.

We now proceed to calculate the second moment,

$$
\langle x^2 \rangle = \int_{-\infty}^{+\infty} x^2 \, P(x, T; x_0, 0) \, dx
$$

$$
= \left(2\pi \int_0^T [f(T-t) \, h(t)]^2 \, dt \right)^{-\frac{1}{2}}
$$

$$
\times \int_{-\infty}^{+\infty} x^2 \, \exp\left(-\left[\int_0^T [f(T-t) \, h(t)]^2 \, dt \right]^{-1} \frac{(x-x_0)^2}{2} \right) dx.
$$

$$(3.31)$$

With Eqs. (3.25) and (3.26), we have,

$$\langle x^2 \rangle = \frac{\alpha}{\sqrt{\pi}} \int\limits_{-\infty}^{+\infty} (y + x_0)^2 \exp\left(-\alpha^2 y^2\right) \, dy$$

$$= \frac{\alpha}{\sqrt{\pi}} \int\limits_{-\infty}^{+\infty} \left(y^2 + 2yx_0 + x_0^2\right) \exp\left(-\alpha^2 y^2\right) \, dy$$

$$= \frac{\alpha}{\sqrt{\pi}} \int\limits_{-\infty}^{+\infty} y^2 \exp\left(-\alpha^2 y^2\right) \, dy + \frac{2\alpha x_0}{\sqrt{\pi}} \int\limits_{-\infty}^{+\infty} y \, \exp\left(-\alpha^2 y^2\right) \, dy$$

$$+ \frac{\alpha x_0^2}{\sqrt{\pi}} \int\limits_{-\infty}^{+\infty} \exp\left(-\alpha^2 y^2\right) \, dy$$

$$= \frac{\alpha}{\sqrt{\pi}} \int\limits_{-\infty}^{+\infty} y^2 \exp\left(-\alpha^2 y^2\right) \, dy + 0 + \frac{\alpha x_0^2}{\sqrt{\pi}} \frac{\sqrt{\pi}}{\alpha} , \tag{3.32}$$

where the 2nd and 3rd integrals are the same as in Eq. (3.30). For the remaining integral, we can use Eq. (3.462.8) of [98] to get,

$$\langle x^2 \rangle = \frac{\alpha}{\sqrt{\pi}} \frac{1}{2\alpha^2} \sqrt{\frac{\pi}{\alpha^2}} + x_0^2$$

$$= x_0^2 + \frac{1}{2\alpha^2}$$

$$= x_0^2 + \int\limits_0^T \left[f\left(T - t\right) h\left(t\right)\right]^2 \, dt , \tag{3.33}$$

where $f\left(T - t\right)$ is the memory function.

We can also obtain the variance $Var\left[x\right]$ which gives us a measure of the width of the distribution. The variance $Var\left[x\right]$ is given by,

$$Var\left[x\right] = \left\langle \left(x - \langle x \rangle\right)^2 \right\rangle$$

$$= \left\langle \left(x^2 - 2x \langle x \rangle + \langle x \rangle^2\right) \right\rangle$$

$$= \langle x^2 \rangle - \langle x \rangle^2$$

$$= \int\limits_0^T \left[f\left(T - t\right) h\left(t\right)\right]^2 \, dt , \tag{3.34}$$

where we used Eqs. (3.30) and (3.33).

3.8   Standard Deviation with Memory Function

Additional insights about displacements along x, on the average, can be obtained. In particular, we can look at the standard deviation σ which measures the degree of deviation of x from the mean value $\langle x \rangle$ [90; 169]. This standard deviation is obtained as the square root of the variance $Var\,[x]$, Eq. (3.34), and is given by the expression (let, $\langle x \rangle = x_0$),

$$\sigma = \sqrt{\left\langle (x - x_0)^2 \right\rangle} = \sqrt{\int_0^T \left[f\,(T - t)\,h\,(t) \right]^2 dt}\,. \tag{3.35}$$

Let us consider some examples of standard deviations involving memory functions using Eq. (3.35).

(1) **Ordinary Brownian Motion**

When the memory function $f\,(T - t)$ is simply a constant κ and $h\,(t) = 1$, we note that Eq. (1.13) reduces to Eq. (1.16) which describes ordinary Brownian motion. In this case, Eq. (3.35) yields the standard deviation,

$$\sigma = \kappa T^{1/2}, \tag{3.36}$$

where T is the final time. Equation (3.36) means that the width of the probability distribution increases with time as $\sim T^{1/2}$, as expected for ordinary Brownian motion.

(2) **Fractional Brownian Motion**

For the memory function $f\,(\tau - t)$ given by Eq. (3.14) associated with the Riemann-Liouville fractional Brownian motion [156], Eq. (3.16), with $h\,(t) = 1$ we need to evaluate the integral in σ of the form,

$$\int_0^T f\,(T - t)^2\,dt = \frac{1}{\Gamma\left(H + \frac{1}{2}\right)^2} \int_0^T (T - t)^{2H-1}\,dt\,, \tag{3.37}$$

where $\Gamma\,(\nu)$ is the Gamma function. If we let, $z = T - t$, and, $\beta = 2H - 1$, the integration in Eq. (3.37) acquires the form,

$$\int_0^T z^\beta\,dz = \frac{T^{\beta+1}}{\beta + 1}\,, \tag{3.38}$$

such that Eq. (3.37) becomes,

$$\int_0^T f\,(T - t)^2\,dt = \frac{T^{2H}}{2H\,\Gamma\left(H + \frac{1}{2}\right)^2}\,. \tag{3.39}$$

Taking the square root of Eq. (3.39) gives the standard deviation Eq. (3.35) for fractional Brownian motion,

$$\sigma = \frac{1}{\sqrt{2H}\,\Gamma\left(H + \frac{1}{2}\right)} T^H, \qquad (3.40)$$

where H is the Hurst exponent with values $0 < H < 1$. For $H = 1/2$, we obtain the time dependence of ordinary Brownian motion, Eq. (3.36). The range $1/2 < H < 1$ describes the enhanced diffusion exhibiting long-memory property, and $0 < H < 1/2$ describes suppressed diffusion for short-memory processes.

(3) **Bessel-modified Brownian Motion**
Let us next look at the oscillatory memory function given by Eq. (3.17). To obtain σ we consider the integral,

$$\int_0^T [f(T-t)\,h(t)]^2\,dt = \int_0^T \sin(T-t)\,J_0(t)\,dt$$

$$= TJ_1(T) \qquad (3.41)$$

where we used Eq. (6.674.7) of [98]. Equations (3.35) and (3.41) yield a standard deviation of the form,

$$\sigma = \sqrt{TJ_1(T)}. \qquad (3.42)$$

The standard deviation σ, therefore, depends on time T modified by an oscillatory Bessel function.

(4) **Exponentially-modified Brownian Motion**
Let us now consider the memory function $f(\tau - t)\,h(t)$ given by Eq. (3.20). The integral in σ, Eq. (3.35), appears as,

$$\int_0^T [f(T-t)\,h(t)]^2\,dt = \int_0^T \frac{(T-t)^{\mu-1}\,e^{-\beta/t}}{t^{\mu+1}}\,dt$$

$$= \frac{\Gamma(\mu)\,T^{\mu-1}e^{-\beta/T}}{\beta^\mu}, \qquad (3.43)$$

where we used Eq. (3.471.3) of [98], with Re $\mu > 0$ and $T > 0$. Equation (3.43), together with Eq. (3.35), gives a standard deviation of the form,

$$\sigma = \sqrt{\frac{\Gamma(\mu)}{\beta^\mu}} T^{(\mu-1)/2} e^{-\beta/2T}. \qquad (3.44)$$

The standard deviation for this case can be dominated by an exponential term $e^{-(\beta/2)T}$ as time T increases.

3.9   Modified Diffusion Equation

We now look at the kinetic equation satisfied by the probability density function $P(x_T, T; x_0, 0)$ with memory function $f(\tau - t)$, given by Eq. (3.12). Using the notation, $x_T = x$, and $T = \tau$, it is straightforward to show that $P(x, \tau; x_0, 0)$ satisfies the equation,

$$\frac{\partial}{\partial \tau} P(x, \tau; x_0, 0) = \left[\frac{1}{2} \frac{\partial}{\partial \tau} \int_0^\tau \left[f(\tau - t) h(t) \right]^2 dt \right] \frac{\partial^2}{\partial x^2} P(x, \tau; x_0, 0),$$

(3.45)

where, instead of a constant diffusion coefficient, a time-dependent diffusive behavior is allowed [207]. One way of verifying Eq. (3.45) is by taking the time derivative of Eq. (3.12), i.e.,

$$\frac{\partial}{\partial \tau} P(x, \tau; x_0, 0) = \frac{\partial}{\partial \tau} \left\{ \frac{1}{\sqrt{2\pi \int_0^\tau \left[f(\tau - t) h(t) \right]^2 dt}} \right.$$

$$\left. \times \exp \left[\frac{-(x - x_0)^2}{2 \int_0^\tau \left[f(\tau - t) h(t) \right]^2 dt} \right] \right\},$$

(3.46)

which yields the expression,

$$\frac{\partial}{\partial \tau} P(x, \tau; x_0, 0) = \left[\frac{1}{2} \frac{\partial}{\partial \tau} \int_0^\tau \left[f(\tau - t) h(t) \right]^2 dt \right]$$

$$\times \left\{ \frac{-P(x, \tau; x_0, 0)}{\int_0^\tau \left[f(\tau - t) h(t) \right]^2 dt} \left[1 - \frac{(x - x_0)^2}{\int_0^\tau \left[f(\tau - t) h(t) \right]^2 dt} \right] \right\}.$$

(3.47)

With Eq. (3.12) an evaluation of, $\left(\partial^2 / \partial x^2 \right) P(x, \tau; x_0, 0)$, however, shows that it is equal to the factor in curly brackets in Eq. (3.47), in particular,

$$\frac{\partial^2}{\partial x^2} P(x, \tau; x_0, 0) = \frac{-P(x, \tau; x_0, 0)}{\int_0^\tau \left[f(\tau - t) h(t) \right]^2 dt} \left[1 - \frac{(x - x_0)^2}{\int_0^\tau \left[f(\tau - t) h(t) \right]^2 dt} \right].$$

(3.48)

Using Eq. (3.48) on the right-hand side of Eq. (3.47) yields an expression for the modified diffusion equation (3.45). We now look at specific examples.

(1) **Ordinary Brownian Motion**

Consider the special case where $f(T - t)$ is just a constant given by $f = \sqrt{2D}$ and $h(t) = 1$. Equation (3.45) yields,

$$\frac{\partial}{\partial \tau} P(x, \tau; x_0, 0) = D \frac{\partial^2}{\partial x^2} P(x, \tau; x_0, 0) \tag{3.49}$$

which is the usual diffusion equation for the Wiener process where D is a diffusion coefficient [147].

(2) **Fractional Brownian Motion**

For a memory function given by Eq. (3.14) and $h(t) = 1$, we have (see, also, Eq. (3.39)),

$$\frac{\partial}{\partial \tau} \int_0^\tau [f(\tau - t) h(t)]^2 \, dt = \frac{\partial}{\partial \tau} \left(\frac{\tau^{2H}}{2H \, \Gamma \left(H + \frac{1}{2} \right)^2} \right)$$

$$= \frac{\tau^{2H-1}}{\Gamma \left(H + \frac{1}{2} \right)^2}. \tag{3.50}$$

Using this in Eq. (3.45) yields,

$$\frac{\partial}{\partial \tau} P(x, \tau; x_0, 0) = \frac{\tau^{2H-1}}{2\Gamma \left(H + \frac{1}{2} \right)^2} \frac{\partial^2}{\partial x^2} P(x, \tau; x_0, 0) \tag{3.51}$$

which is the modified diffusion equation for fractional Brownian motion in Riemann-Liouville representation [46].

(3) **Bessel-modified Brownian Motion**

From Eq. (3.41) for the Bessel-modified Brownian motion, we have,

$$\frac{\partial}{\partial \tau} \int_0^\tau f(\tau - t)^2 \, dt = \frac{\partial}{\partial \tau} [\tau J_1(\tau)]$$

$$= J_1(\tau) + \tau \frac{\partial}{\partial \tau} [J_1(\tau)]$$

$$= J_1(\tau) + \tau [J_0(\tau) - J_2(\tau)] \tag{3.52}$$

where we used the recurrence relation (combination of Eqs. (8.472.1) and (8.472.2) of [98]),

$$\frac{\partial}{\partial x} J_\nu(x) = \frac{1}{2} J_{\nu-1}(x) - \frac{1}{2} J_{\nu+1}(x). \tag{3.53}$$

This yields, from Eq. (3.45), a diffusion equation of the form,

$$\frac{\partial}{\partial \tau} P(x, \tau; x_0, 0) = \frac{1}{4} [J_{\nu-1}(x) - J_{\nu+1}(x)] \frac{\partial^2}{\partial x^2} P(x, \tau; x_0, 0). \tag{3.54}$$

(4) **Exponentially-modified Brownian Motion**

For the memory function $f(\tau - t)$ and $h(t)$ given by Eq. (3.20), we have (see, also, Eq. (3.43)),

$$\frac{\partial}{\partial \tau} \int_0^\tau [f(\tau - t) h(t)]^2 \, dt = \frac{\partial}{\partial \tau} \left(\frac{\Gamma(\mu) \tau^{\mu-1} e^{-\beta/\tau}}{\beta^\mu} \right)$$

$$= \frac{\Gamma(\mu) e^{-\beta/\tau}}{\beta^\mu} \left[(\mu - 1) \tau^{\mu-2} + \beta \tau^{\mu-3} \right].$$

(3.55)

This gives rise to a modified diffusion equation of the form,

$$\frac{\partial}{\partial \tau} P(x, \tau; x_0, 0) = \left(\frac{\Gamma(\mu) e^{-\beta/\tau}}{2\beta^\mu} \left[(\mu - 1) \tau^{\mu-2} + \beta \tau^{\mu-3} \right] \right)$$

$$\times \frac{\partial^2}{\partial x^2} P(x, \tau; x_0, 0).$$

(3.56)

3.10   Example: Periodic Boundary Condition

In many systems, a particle could be subjected to a periodic boundary condition such as those in a regular lattice, or spaces with circular topology [185]. For instance, a particle moving to the right or left in an infinite one-dimensional periodic lattice, with lattice spacing d, may also be modelled by a particle which moves clockwise or counterclockwise in a circle with circumference d. As an example, therefore, let us consider diffusion in a circle where the paths are parametrized by an angular variable given by,

$$\varphi(\tau) = \varphi_0 + \int_0^\tau f(\tau - t) \; h(t) \; \omega(t) \, dt,$$

(3.57)

where, $f(\tau - t)$, is a memory function, $dB(t) = \omega(t) \, dt$, and φ_0 is the initial point.

If we are interested in paths which start at φ_0 and end at φ_1 (see Figure 3.2), then we look at all paths which satisfy the constraint,

$$\delta(\varphi(\tau) - \varphi_1) = \delta \left(\varphi_0 - \varphi_1 + \int_0^\tau f(\tau - t) \; h(t) \; \omega(t) \, dt \right).$$

(3.58)

We note that from φ_0, a particle can go clockwise or counterclockwise on a circle to arrive at φ_1. The particle could even diffuse several times around

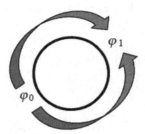

Fig. 3.2.   A particle from an initial point φ_0 can go clockwise or counterclockwise to arrive at a final point φ_1.

the circle before stopping at φ_1. In view of the topologically inequivalent paths, we sum over all possible paths or histories with endpoints, φ_0 and φ_1, i.e.,

$$\sum_{n=-\infty}^{+\infty} \delta\left(\varphi\left(\tau\right) - \varphi_1 + 2\pi n\right) \quad ; \quad n = 0, \pm 1, \pm 2, \dots . \tag{3.59}$$

To get the probability $P\left(\varphi_1, \tau; \varphi_0, 0\right)$ that indeed $\varphi\left(\tau\right)$ ends at φ_1 at time τ, if it started at φ_0, we evaluate the expectation value of the paths expressed by Eq. (3.59), and write,

$$P\left(\varphi_1, \tau; \varphi_0, 0\right) = \mathbb{E}\left(\sum_{n=-\infty}^{+\infty} \delta\left(\varphi\left(\tau\right) - \varphi_1 + 2\pi n\right)\right)$$

$$= \int \left(\sum_{n=-\infty}^{+\infty} \delta\left(\varphi\left(\tau\right) - \varphi_1 + 2\pi n\right)\right) d\mu$$

$$= \sum_{n=-\infty}^{+\infty} \int \delta\left(\varphi\left(\tau\right) - \varphi_1 + 2\pi n\right) d\mu\,, \tag{3.60}$$

where $d\mu$ is the Gaussian white noise measure. We can express the delta function in terms of its Fourier representation to get,

$$P\left(\varphi_1, \tau; \varphi_0, 0\right) = \sum_{n=-\infty}^{+\infty} \int \frac{1}{2\pi} \int_{-\infty}^{+\infty} \exp\left[ik\left(\varphi\left(\tau\right) - \varphi_1 + 2\pi n\right)\right] dk \; d\mu$$

$$= \sum_{n=-\infty}^{+\infty} \int_{-\infty}^{+\infty} \frac{dk}{2\pi} \exp\left[ik\left(\varphi_0 - \varphi_1 + 2\pi n\right)\right]$$

$$\times \int \exp\left[ik \int_0^{\tau} f\left(\tau - t\right) \; h\left(t\right) \; \omega\left(t\right) dt\right] d\mu\,, \tag{3.61}$$

where we used Eq. (3.57) for $\varphi(\tau)$. For the integral over $d\mu$ we use Eq. (2.6), i.e.,

$$\int \exp\left(i \int \omega(t)\,\xi(t)\,dt\right) d\mu(\omega) = \exp\left(-\frac{1}{2}\int \xi^2(t)\,dt\right), \qquad (3.62)$$

for the characteristic functional, where $\xi(t) = k\,f(\tau - t)\,h(t)$. We get,

$$P(\varphi_1, \tau; \varphi_0, 0) = \sum_{n=-\infty}^{+\infty} \int_{-\infty}^{+\infty} \frac{dk}{2\pi} \exp\left[ik(\varphi_0 - \varphi_1 + 2\pi n)\right]$$

$$\times \exp\left(-\frac{k^2}{2}\int_0^\tau \left[f(\tau - t)\,h(t)\right]^2 \, dt\right). \qquad (3.63)$$

The remaining integral over k is a Gaussian integral which could be evaluated to yield,

$$P(\varphi_1, \tau; \varphi_0, 0) = \sum_{n=-\infty}^{+\infty} P_n(\varphi_1, \tau; \varphi_0, 0) \qquad (3.64)$$

where n is a winding number and,

$$P_n(\varphi_1, \tau; \varphi_0, 0) = \left(2\pi \int_0^\tau \left[f(\tau - t)\,h(t)\right]^2 \, dt\right)^{-\frac{1}{2}}$$

$$\times \exp\left(-\frac{(\varphi_0 - \varphi_1 + 2\pi n)^2}{2\int_0^\tau \left[f(\tau - t)\,h(t)\right]^2 \, dt}\right). \qquad (3.65)$$

Alternatively, from Eq. (3.63), we can apply the Poisson sum formula,

$$\frac{1}{2\pi} \sum_{n=-\infty}^{+\infty} \exp(in\alpha) = \sum_{m=-\infty}^{+\infty} \delta(\alpha + 2\pi m) \qquad (3.66)$$

to obtain the expression,

$$P(\varphi_1, \tau; \varphi_0, 0) = \frac{1}{2\pi} \sum_{m=-\infty}^{+\infty} \int_{-\infty}^{+\infty} \delta(k + m) \exp\left[ik(\varphi_0 - \varphi_1)\right]$$

$$\times \exp\left(-\frac{k^2}{2}\int_0^\tau \left[f(\tau - t)\,h(t)\right]^2 \, dt\right) dk. \qquad (3.67)$$

The integration in Eq. (3.67) is facilitated by the delta function to yield,

$$P(\varphi_1, \tau; \varphi_0, 0) = \frac{1}{2\pi} \sum_{m=-\infty}^{+\infty} \exp\left[-im(\varphi_0 - \varphi_1)\right]$$

$$\times \exp\left(-\frac{m^2}{2} \int_0^\tau \left[f(\tau - t) h(t)\right]^2 dt\right). \qquad (3.68)$$

The probability function Eq. (3.68) can be expressed as [26],

$$P(\varphi_1, \tau; \varphi_0, 0) = \frac{1}{2\pi} \sum_{m=-\infty}^{+\infty} q^{m^2} \exp(i2mu)$$

$$P(\varphi_1, \tau; \varphi_0, 0) = \frac{1}{2\pi} \Theta_3(u) \qquad (3.69)$$

where $q = \exp\left[-\frac{1}{2} \int_0^\tau \left[f(\tau - t) h(t)\right]^2 dt\right]$, $\Theta_3(u)$ is a Theta function, and $u = -(\varphi_0 - \varphi_1)/2$. Note that, we could also write,

$$\Theta_3(u) = 1 + 2 \sum_{m=1}^{\infty} q^{m^2} \cos(2mu). \qquad (3.70)$$

From these results, we can evaluate the winding probability W_n from Eqs. (3.65) and (3.69), i.e.,

$$W_n = \frac{P_n(\varphi_1, \tau; \varphi_0, 0)}{P(\varphi_1, \tau; \varphi_0, 0)}$$

$$W_n = \sqrt{\frac{2\pi}{\int_0^\tau \left[f(\tau - t) h(t)\right]^2 dt}} \exp\left(-\frac{(\varphi_0 - \varphi_1 + 2\pi n)^2}{2 \int_0^\tau \left[f(\tau - t) h(t)\right]^2 dt}\right)$$

$$\times \left[\Theta_3\left(\frac{\varphi_0 - \varphi_1}{2}\right)\right]^{-1}. \qquad (3.71)$$

3.11 Example: The Wedge Boundary for Diffusion with Memory

Consider a wedge with an angular opening of $\bar{\varphi}$ (see Figure 3.3). For diffusion of a particle in the space bounded by the wedge, we can again use the parametrization given by Eq. (3.57).

As shown in Figure 3.4a, a particle from φ_0 that goes directly to φ_1 is described by the constraint, $\delta(\vartheta(\tau) - \vartheta_1)$. Alternatively, the particle from

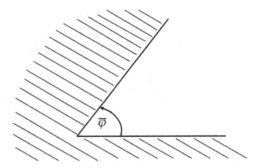

Fig. 3.3.   The angle $\bar{\varphi}$ is a constant measuring the opening of the wedge.

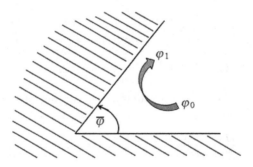

Fig. 3.4a.   A particle from φ_0 can go directly to φ_1.

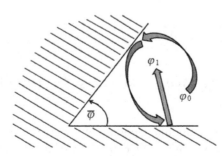

Fig. 3.4b.   A particle from φ_0 can bounce off the walls before ending at φ_1 traveling an extra $2\bar{\varphi}n$ ($n = 0, \pm1, \pm2, ...$) distance.

φ_0 can bounce off the walls before ending at φ_1, thus traveling an extra $2\bar{\varphi}n$ ($n = 0, \pm1, \pm2, ...$) distance (see Figure 3.4b). Taking into account all possible paths that arrive at φ_1 we, therefore, consider a constraint described by the summation,

$$\sum_{n=-\infty}^{+\infty} \delta\left(\varphi\left(\tau\right) - \varphi_1 + 2\bar{\varphi}n\right). \tag{3.72}$$

3.11.1 *Dirichlet Boundary Condition*

Suppose we now wish to have a solution $P_w\left(\varphi_1, \tau; \varphi_0, 0\right)$ for the wedge problem which satisfies the Dirichlet boundary condition where

$$P_w\left(\varphi_1, \tau; \varphi_0, 0\right) = 0 , \qquad \text{at } \varphi = 0 \text{ and } \varphi = \bar{\varphi}.$$

From Eq. (3.72), we look at the linear combination,

$$C_n = \delta\left(\varphi\left(\tau\right) - \varphi_1 + 2\bar{\varphi}n\right) - \delta\left(\varphi\left(\tau\right) + \varphi_1 + 2\bar{\varphi}n\right). \tag{3.73}$$

We can sum over all possible values of n, i.e., $\sum_n C_n$, and evaluate the expectation value for a particle to end at φ_1 if it started at φ_0. We have,

$$P_w\left(\varphi_1, \tau; \varphi_0, 0\right) = \int \left(\sum_{n=-\infty}^{+\infty} C_n\right) d\mu. \tag{3.74}$$

With the substitution of Eq. (3.73), Eq. (3.74) can be written as,

$$P_w\left(\varphi_1, \tau; \varphi_0, 0\right) = \frac{\pi}{\bar{\varphi}} \int \sum_{n=-\infty}^{+\infty} \left[\delta\left(\frac{\pi}{\bar{\varphi}}\left(\varphi\left(\tau\right) - \varphi_1\right) + 2\pi n\right)\right.$$
$$\left. -\delta\left(\frac{\pi}{\bar{\varphi}}\left(\varphi\left(\tau\right) + \varphi_1\right) + 2\pi n\right)\right] d\mu \tag{3.75}$$

for which we apply the Poisson sum formula, Eq. (3.66) to get,

$$P_w\left(\varphi_1, \tau; \varphi_0, 0\right) = \frac{1}{2\bar{\varphi}} \int \sum_{m=-\infty}^{+\infty} \left\{\exp\left[\frac{im\pi}{\bar{\varphi}}\left(\varphi\left(\tau\right) - \varphi_1\right)\right]\right.$$
$$\left. - \exp\left[\frac{im\pi}{\bar{\varphi}}\left(\varphi\left(\tau\right) + \varphi_1\right)\right]\right\} d\mu. \tag{3.76}$$

Using Eq. (3.57) for $\varphi(\tau)$, Eq. (3.76) gives,

$$
P_w(\varphi_1, \tau; \varphi_0, 0) = \frac{1}{2\bar{\varphi}} \int \sum_{m=-\infty}^{+\infty} \left\{ \exp\left[\frac{im\pi}{\bar{\varphi}} (\varphi_0 - \varphi_1) \right] \right.
$$

$$
- \exp\left[\frac{im\pi}{\bar{\varphi}} (\varphi_0 + \varphi_1) \right] \Bigg\}
$$

$$
\times \exp\left(\frac{im\pi}{\bar{\varphi}} \int_0^\tau f(\tau - t) h(t) \omega(t)\, dt \right) d\mu .
$$

$$
(3.77)
$$

We can then rewrite the exponential terms as,

$$
\sum_{m=-\infty}^{+\infty} \left\{ \exp\left[\frac{im\pi}{\bar{\varphi}} (\varphi_0 - \varphi_1) \right] - \exp\left[\frac{im\pi}{\bar{\varphi}} (\varphi_0 + \varphi_1) \right] \right\}
$$

$$
= 2 \sum_{m=-\infty}^{+\infty} \sin\left(\frac{m\pi}{\bar{\varphi}} \varphi_0 \right) \sin\left(\frac{m\pi}{\bar{\varphi}} \varphi_1 \right), \qquad (3.78)
$$

to obtain for Eq. (3.77), the form,

$$
P_w(\varphi_1, \tau; \varphi_0, 0) = \frac{1}{\bar{\varphi}} \sum_{m=-\infty}^{+\infty} \sin\left(\frac{m\pi}{\bar{\varphi}} \varphi_0 \right) \sin\left(\frac{m\pi}{\bar{\varphi}} \varphi_1 \right)
$$

$$
\times \int \exp\left(\frac{im\pi}{\bar{\varphi}} \int_0^\tau f(\tau - t) h(t) \omega(t)\, dt \right) d\mu .
$$

$$
(3.79)
$$

The integral over the white noise measure $d\mu$ can be carried out using Eq. (3.62) where $\xi(t) = (m\pi/\bar{\varphi})\, f(\tau - t) h(t)$ to yield,

$$
P_w(\varphi_1, \tau; \varphi_0, 0) = \frac{1}{\bar{\varphi}} \sum_{m=-\infty}^{+\infty} \sin\left(\frac{m\pi}{\bar{\varphi}} \varphi_0 \right) \sin\left(\frac{m\pi}{\bar{\varphi}} \varphi_1 \right)
$$

$$
\times \exp\left[-\frac{m^2\pi^2}{2\bar{\varphi}^2} \int_0^\tau [f(\tau - t) h(t)]^2\, dt \right], \quad (3.80)
$$

or,

$$P_w\left(\varphi_1, \tau; \varphi_0, 0\right) = \sum_{m=-\infty}^{+\infty} \Phi_m\left(\varphi_0\right) \Phi_m\left(\varphi_1\right)$$

$$\times \exp\left[-\frac{m^2\pi^2}{2\bar{\varphi}^2} \int_0^\tau \left[f\left(\tau - t\right) h\left(t\right)\right]^2 dt\right], \quad (3.81)$$

with, $\Phi_m\left(\varphi\right) = (1/\sqrt{\bar{\varphi}}) \sin\left(m\pi\varphi/\bar{\varphi}\right)$, which vanishes at $\varphi = 0$ and $\varphi = \bar{\varphi}$, or the walls of the wedge satisfying the Dirichlet boundary condition. As shown in Table 3.1, one could then consider different forms of the memory function $f\left(\tau - t\right)$ and $h\left(t\right)$ to model specific situations.

3.11.2 *Neumann Boundary Condition*

We can also look for solutions which satisfy the Neumann boundary condition,

$$\frac{\partial}{\partial \varphi_1} P_w'\left(\varphi_1, \tau; \varphi_0, 0\right) = 0 , \qquad \text{at} \qquad \varphi_1 = \bar{\varphi} \text{ or } \varphi_1 = 0 . \quad (3.82)$$

Using again the form of Eq. (3.72), we look at the combination,

$$C_n' = \delta\left(\varphi\left(\tau\right) - \varphi_1 + 2\bar{\varphi}n\right) + \delta\left(\varphi\left(\tau\right) + \varphi_1 + 2\bar{\varphi}n\right). \quad (3.83)$$

Summing over all possible values of n, i.e., $\sum_n C_n'$, we evaluate the expectation value for a particle to end at φ_1 if it started at φ_0, i.e.,

$$P_w'\left(\varphi_1, \tau; \varphi_0, 0\right) = \int \left(\sum_{n=-\infty}^{+\infty} C_n'\right) d\mu . \quad (3.84)$$

Equation (3.84) has the form,

$$P_w'\left(\varphi_1, \tau; \varphi_0, 0\right) = \frac{\pi}{\bar{\varphi}} \int \sum_{n=-\infty}^{+\infty} \left[\delta\left(\frac{\pi}{\bar{\varphi}}\left(\varphi\left(\tau\right) - \varphi_1\right) + 2\pi n\right)\right.$$

$$\left. + \delta\left(\frac{\pi}{\bar{\varphi}}\left(\varphi\left(\tau\right) + \varphi_1\right) + 2\pi n\right)\right] d\mu , \quad (3.85)$$

where we can apply the Poisson sum formula, Eq. (3.66) to obtain,

$$P'_w(\varphi_1, \tau; \varphi_0, 0) = \frac{1}{2\bar{\varphi}} \int \sum_{m=-\infty}^{+\infty} \left\{ \exp\left[\frac{im\pi}{\bar{\varphi}}\left(\varphi(\tau) - \varphi_1\right)\right] \right.$$

$$\left. + \exp\left[\frac{im\pi}{\bar{\varphi}}\left(\varphi(\tau) + \varphi_1\right)\right] \right\} d\mu. \quad (3.86)$$

With Eq. (3.57) for $\varphi(\tau)$, Eq. (3.86) becomes,

$$P'_w(\varphi_1, \tau; \varphi_0, 0) = \frac{1}{2\bar{\varphi}} \int \sum_{m=-\infty}^{+\infty} \left\{ \exp\left[\frac{im\pi}{\bar{\varphi}}\left(\varphi_0 - \varphi_1\right)\right] \right.$$

$$\left. + \exp\left[\frac{im\pi}{\bar{\varphi}}\left(\varphi_0 + \varphi_1\right)\right] \right\}$$

$$\times \exp\left(\frac{im\pi}{\bar{\varphi}} \int_0^\tau f(\tau - t) h(t) \omega(t) dt\right) d\mu.$$

$$(3.87)$$

The exponential terms can be expressed as,

$$\sum_{m=-\infty}^{+\infty} \left\{ \exp\left[\frac{im\pi}{\bar{\varphi}}\left(\varphi_0 - \varphi_1\right)\right] + \exp\left[\frac{im\pi}{\bar{\varphi}}\left(\varphi_0 + \varphi_1\right)\right] \right\}$$

$$= 2 \sum_{m=-\infty}^{+\infty} \cos\left(\frac{m\pi}{\bar{\varphi}}\varphi_0\right) \cos\left(\frac{m\pi}{\bar{\varphi}}\varphi_1\right), \quad (3.88)$$

to obtain for Eq. (3.87) the form,

$$P'_w(\varphi_1, \tau; \varphi_0, 0) = \frac{1}{\bar{\varphi}} \sum_{m=-\infty}^{+\infty} \cos\left(\frac{m\pi}{\bar{\varphi}}\varphi_0\right) \cos\left(\frac{m\pi}{\bar{\varphi}}\varphi_1\right)$$

$$\times \int \exp\left(\frac{im\pi}{\bar{\varphi}} \int_0^\tau f(\tau - t) h(t) \omega(t) \, dt\right) d\mu.$$

$$(3.89)$$

The integral over the white noise measure $d\mu$ is done using Eq. (3.62) where $\xi(t) = (m\pi/\bar{\varphi})\ f(\tau - t)\,h(t)$. Equation (3.89) then gives,

$$P'_w(\varphi_1, \tau; \varphi_0, 0) = \frac{1}{\bar{\varphi}} \sum_{m=-\infty}^{+\infty} \cos\left(\frac{m\pi}{\bar{\varphi}}\varphi_0\right) \cos\left(\frac{m\pi}{\bar{\varphi}}\varphi_1\right)$$

$$\times \exp\left[-\frac{m^2\pi^2}{2\bar{\varphi}^2} \int_0^\tau [f(\tau - t)\,h(t)]^2\,dt\right], \quad (3.90)$$

or

$$P'_w(\varphi_1, \tau; \varphi_0, 0) = \sum_{m=-\infty}^{+\infty} \Phi_m(\varphi_0)\ \Phi_m(\varphi_1)$$

$$\times \exp\left[-\frac{m^2\pi^2}{2\bar{\varphi}^2} \int_0^\tau [f(\tau - t)\,h(t)]^2\,dt\right], \quad (3.91)$$

where $\Phi_m(\varphi) = (1/\sqrt{\bar{\varphi}})\cos(m\pi\varphi/\bar{\varphi})$. Some possible memory functions $f(\tau - t)$ and $h(t)$ may then be considered from Table 3.1.

Exercises

(3-1) For fluctuations with oscillatory memory function described by,

$$f(T - t)\ h(t) = \cos^{\frac{1}{2}}(T - t)\ \sqrt{J_0(t)},$$

where $J_v(t)$ is a Bessel function of the first kind:
  (a) Obtain the probability density function $P(x_T, T; x_0, 0)$.
  (b) Derive the modified diffusion equation satisfied by $P(x_T, T; x_0, 0)$.

(3-2) Show that Eq. (3.91) satisfies the Neumann boundary condition shown in Eq. (3.82).

Chapter 4

Complex Systems

Investigation of complex systems with non-Markovian or memory properties may be aided by the approach presented here. Complex systems generally have many interrelated components that give rise to an observable collective behavior or property. Activities of human societies, for example, often exhibit complexity that can lead to economic growth. Other examples would be climate changes, transportation systems, telecommunication infrastructure, and biological immune systems. Although interactions of different components can be intractable, a macroscopic observable in a complex system may exhibit some regularity in behavior. Hence, in spite of not knowing detailed descriptions of its interacting parts, the evolution of a complex system can be modelled. The key step is to parametrize the characteristic observable quantity using Eq. (1.13).

For a given complex system, let us consider a macroscopic quantity, labelled S, which fluctuates in time. To track the decrease or increase of S we define its growth rate as, $R = S_1/S_0$, where S_0 and S_1 are values determined in two consecutive periods. In many cases, it would be convenient to define the logarithm, $x = \ln R = \ln(S_1/S_0)$. Furthermore, a universal feature among complex systems can be illustrated using the memory function Eq. (3.14) with $h(t) = 1$. In this case, the path parametrization, Eq. (1.13), becomes,

$$x(\tau) = x_0 + B^H, \qquad (4.1)$$

where B^H is the fractional Brownian motion, Eq. (3.16), with the corresponding probability density function given by Eq. (3.15).

4.1 Scaling Property

With Eq. (4.1) where $x = \ln(S_1/S_0)$, the probability distribution function $P(x_T, T; x_0, 0)$ of the form given by Eq. (3.15) is known to possess a self-similarity property [144]. We review this feature since it will facilitate our investigation of scaling that manifests in complex systems. We first simplify Eq. (3.15) by defining, $\rho = x_0 - x_T$, and $D = 1/4H\Gamma^2(H + 1/2)$, and writing $P(x_T, T; x_0, 0) = P(\rho, T)$ as,

$$P(\rho, T) = \sqrt{\frac{1}{4\pi DT^{2H}}} \exp\left\{-\frac{\rho^2}{4DT^{2H}}\right\}. \tag{4.2}$$

If we apply the scale transformation,

$$\rho \to \kappa^{-1}\rho \quad ; \quad T \to \kappa^{-1/H}T \tag{4.3}$$

for a given constant κ, we obtain

$$P\left(\kappa^{-1}\rho, \kappa^{-1/H}T\right) = \kappa\, P(\rho, T). \tag{4.4}$$

Equation (4.4) shows self-similar behavior of the probability distribution function.

The Laplace transform $\mathcal{L}[\cdot]$ of the scaled probability distribution function $P\left(\kappa^{-1}\rho, \kappa^{-1/H}T\right)$, Eq. (4.4), can also be evaluated. We have,

$$\mathcal{L}\left[P\left(\kappa^{-1}\rho, \kappa^{-1/H}T\right)\right] = \frac{1}{2}\sqrt{\frac{\kappa}{D}} \exp\left\{-\sqrt{\frac{\kappa}{D}}\,|\rho|\right\}. \tag{4.5}$$

In Eq. (4.5), the width of the probability distribution, or standard deviation σ is identified as, $\sigma = \sqrt{2D/\kappa}$. Designating the initial and final points of the fractional Brownian motion as, $x_0 = \bar{r}$ and $x_T = r$, respectively, where $\bar{r}$ is the average growth rate, and writing $|\rho| = |r - \bar{r}|$, then $P(r) = \mathcal{L}\left[P\left(\kappa^{-1}\rho, \kappa^{-1/H}T\right)\right]$, Eq. (4.5), acquires the form [34],

$$P(r) = \frac{1}{\sqrt{2}\sigma} \exp\left(-\frac{\sqrt{2}\,|r - \bar{r}|}{\sigma}\right). \tag{4.6}$$

From Eq. (4.6), a plot of $P(r)$ versus r gives a tent-shaped graph as shown in Fig. 4.1. The time-independent Laplace distribution, Eq. (4.6), has exactly the same form as the empirical formula derived in earlier studies of complex systems [8; 41; 136; 171; 172; 194]. We discuss some examples in the following sections.

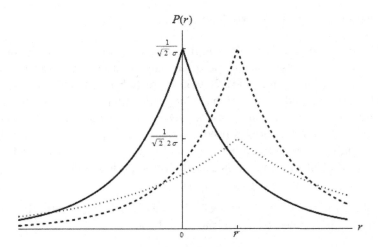

Fig. 4.1. Exponential Laplace distribution: $\bar{r} = 0$ (bold); $\bar{r} \neq 0$ (dashed). Doubling σ reduces the height of the distribution (dotted).

4.2 Metabolic Rate Fluctuation

The rate at which an animal consumes oxygen is referred to as the metabolic rate (VO_2). This scales with the body mass. Since the rate of oxygen consumption can fluctuate, the metabolic rate fluctuations within individual organisms was examined for 71 individuals belonging to 25 species of small terrestrial vertebrates (10 bird species, 12 small mammals, and 3 lizards) [136]. The macroscopic variable S under consideration is, therefore, $S = VO_2$, and the expression, $x = \ln(S_1/S_0)$ discussed earlier translates into the growth rate of VO_2, i.e., $x \equiv \log[VO_2(t + \tau)/VO_2(t)]$ where $VO_2(t)$ and $VO_2(t + \tau)$ are the metabolic rates observed for each species in time intervals t and $t + \tau$.

By studying the VO_2 time series, F. Labra's group experimentally showed that the distributions of VO_2 fluctuations of individual organisms obey a simple tent-shaped distribution as given by Figure 1 of their paper [136]. In particular, they obtained the Laplace distribution given by Eq. (4.6) where $\bar{r}$ and σ correspond to the mean and standard deviation of VO_2 growth rates, respectively (see also Figure 4.1). The experiment, therefore, agrees with the analytical result embodied in Eq. (4.6) obtained by incorporating memory into the model.

4.3 Fluctuations in Word Use

Another example of a dynamic complex system is language which consists of components (words) and people interacting as users. The properties of 10^7 words recorded from 1800 to 2008 in English, Spanish and Hebrew were recently analyzed and showed to have language independent patterns [171]. The huge book digitization effort of *Google Inc.* involving a database of words in seven languages was used as source of empirical language data. Designating $u_i(t)$ as the number of uses of word i in year t, the fraction of uses of word i out of all word uses in the same period is defined as,

$$f_i(t) \equiv u_i(t) \ / \ N_u(t) \tag{4.7}$$

where $N_u(t)$ is the total number of indistinct word uses printed in year t. The authors proceeded to define the growth rate as,

$$r_i(t) \equiv \ln\left(\frac{f_i(t+\tau)}{f_i(t)}\right) \tag{4.8}$$

for single year growth rates $\tau = 1$. In considering the characteristic time for a word's general acceptance, the $r_i(t)$ was then normalized as, $r_i'(\tau) \equiv r_i(\tau)/\sigma[r_i]$, where $\sigma[r_i]$ is the growth rate standard deviation and τ the age for each new word i. The probability density function $P(R)$ of $R \equiv r_i'(\tau)/\sigma[r'(\tau \mid T_c)]$ was evaluated where time T_c is a threshold to distinguish words that were born in different historical eras. The empirical probability density function shown in Figure 6 of the paper by Petersen *et al.* [171] showed a tent-shaped distribution of the form (using the notation of Petersen *et al.*),

$$P(R) \equiv \frac{1}{\sqrt{2}\sigma(R)} \exp\left[-\frac{\sqrt{2}\,|R - \langle R \rangle|}{\sigma(R)}\right], \tag{4.9}$$

where $\langle R \rangle$ is the average growth rate and $\sigma(R)$ the standard deviation. Clearly, Eq. (4.9) is of the same form as Eq. (4.6) obtained using fractional Brownian motion.

4.4 Growth of Companies

Our last example deals with annual growths based on data of all publicly traded US manufacturing companies producing products of all kinds between 1975 and 1991 [194]. Data were taken from Compustat database where all values for sales were adjusted to 1987 dollars by the GNP price deflator.

For this case, the macroscopic quantity S is identified as the sales of a company. A firm's annual growth rate is then defined as, $r = \ln(S_1/S_0)$, based on the sales S_0 and S_1 for two consecutive years. Evaluating the average growth rate $\bar{r}$ and standard deviation σ from a data set where companies have the same initial sale S_0, the empirically-based work of Stanley *et al.* [194] yields a tent-shaped graph obeying the distribution function, Eq. (4.6).

The present formalism can also accommodate the feature of universality found by Stanley *et al.* [194] showing that scaling holds such that data for a wide range of parameter values collapse into a universal curve. In this case, the distribution function $P(r)$ becomes independent of the initial sales value S_0. To obtain a universal curve, we apply the same scale transformation as Eq. (4.3), but using 2κ instead of just κ where Eq. (4.4) acquires a factor 2. Setting $\kappa = D$, the Laplace transform of the rescaled probability distribution $P(r) = \mathcal{L}\left[P\left((2\kappa)^{-1}x, (2\kappa)^{-1/H}T\right)\right]$ acquires the form,

$$P(r) = \exp\{-|x|\}. \qquad (4.10)$$

Equation (4.10) then corresponds to the scaled probability density $p_{\text{scal}} = \exp(-|r_{\text{scal}}|)$ of Stanley *et al.* [194] where now all data regardless of the initial value of S_0 collapse upon a single curve.

Evidence for the tent-shaped growth rate distribution include Italian manufacturing companies [41], as well as that shown by Alfarano and Milaković [8] where the Subbotin family of distributions were considered. Guided by data from the *Forbes Global 2000* list of the world's largest companies, they showed that the shape parameter α is close to unity such that the Laplace distribution best agrees with the dataset. In particular Eq. (4.6) was obtained [8], differing only by a factor of $\sqrt{2}$ for σ.

4.5 Sensitivity to Changes in Hurst Index

In complex systems, there is an observed scaling relation between the standard deviation of growth rates σ and the quantity S whose growth fluctuation is being studied. In particular, one has [136; 171; 194],

$$\sigma = aS_0^{-\beta} \qquad (4.11)$$

where a is a constant (note that β in [136] is defined as $-\beta$ in [194] and [171]). From Section 4.1, however, we have $\sigma = \sqrt{2D/\kappa}$, where $D = 1/4H\Gamma^2(H+1/2)$. Thus, an analytical form for the standard deviation is obtained, i.e., $\sigma = 1/\sqrt{2\kappa H}\,\Gamma(H+1/2)$. From this and Eq. (4.11), we get an expression for the initial value S_0 in terms of the Hurst index H, i.e. [34],

$$S_0 = \left[a\ \Gamma\left(H + \frac{1}{2} \right) (2\kappa H)^{1/2} \right]^{1/\beta}. \tag{4.12}$$

Since short- and long-term memory are characterized by the Hurst index H, we can look at changes in the quantity S_0 as H goes to a higher value H'. In particular, we define an improvement factor, $\gamma = H'/H \geq 1$. The graph of S_0'/S_0 versus H for different values of γ is given in Figure 4.2. Clearly, the straight horizontal line at the bottom of Figure 4.2 indicates that when $H = H'$ or $\gamma = 1$, then $S_0' = S_0$, whatever the initial value of H. The effect of increasing the improvement factor, $\gamma = H'/H > 1$, up to $\gamma = 2$, shows a dramatic increase in the corresponding S_0' relative to S_0, especially if the initial H is below $H = 0.2$ (short-memory domain).

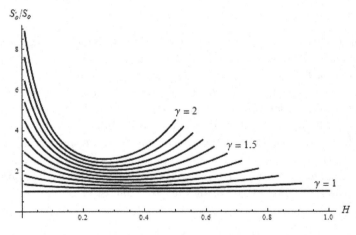

Fig. 4.2.   Increase in fluctuating variable S for different improvement factors $\gamma = H'/H$.

Chapter 5

Time Series Analysis

When we try to understand a given phenomenon, a certain parameter or variable is normally identified and its evolution in time closely monitored through a series of recordings referred to as a time series. In many natural and social phenomena, a time series has variables appearing to fluctuate randomly. Examples of these are stock prices and various financial time series [9; 12; 48; 177; 180], as well as datasets from network and internet traffic [117; 143], complex heartbeat dynamics [10], neurophysiological data from functional magnetic resonance imaging (fMRI) [45], sunspot number fluctuations [158], records of precipitation and river water levels [116; 109; 110] (see Figure 5.1), among many others. It is possible, however, that the multiple factors and interrelated components affecting the evolution of a time series can give rise to an observable collective behavior or property. More importantly, as in other complex systems, a variable may exhibit some universality in behavior which enables us to have a predictive model, in spite of not knowing detailed descriptions of interacting components. Many works exist on the statistical, numerical and theoretical methods dealing with time series [43; 44; 212]. This chapter, however, focuses investigation from the perspective that an underlying dynamic structure of a time series may possess various types of memory behavior. This could give insights into correlations of variables of interest with past values and could increase the feasibility of forecasting future values in the series.

In the following sections, we analyze fluctuations in a time series as stochastic processes with memory properties. Given a fluctuating variable, the mean square displacement of the fluctuation is then examined. We also discuss, as examples, large scale phenomena like typhoon track fluctuations and small scale events such as diffusion of microparticles in complex fluids often investigated in microrheology.

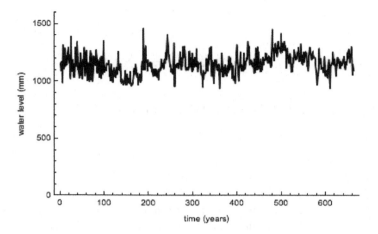

Fig. 5.1. Recordings of the changing water levels of the River Nile.

5.1   Time Series Fluctuation as Modified Brownian Motion

In modelling a fluctuating quantity $x(\tau)$ which starts at an initial value x_0, we parametrize it as (see Eq. (3.6)),

$$x(T) = x_0 + \int_0^T f(T - t)\ h(t)\, dB(t).\tag{5.1}$$

Here, $f(T - t)$ modifies ordinary Brownian motion $B(t)$ at each step as the system progresses from an initial time 0 to a given time T. As earlier shown, the $f(T - t)$ which embodies and determines the type of changes to $B(t)$ can be viewed as a memory function.

Let us now designate $P(x_T, T; x_0, 0)$ as the probability density function giving the probability that at a later time, $t = T$, the fluctuation ends at a specific point $x(T) = x_T$. The situation can be described by an ensemble of all possible paths which start at x_0 and, following Feynman's sum-over-all possible histories [25; 46; 86], which satisfy the pinning constraint, $\delta(x(T) - x_T)$, where $x(T)$ is given by Eq. (5.1). Expressing $dB(t) = \omega(t)\, dt$, where $\omega(t)$ is a Gaussian white noise variable, the delta function constraint appears as,

$$\delta(x(T) - x_T) = \delta\left(x_0 + \int_0^T f(T - t)h(t)\omega(t)\, dt - x_T\right).\tag{5.2}$$

The conditional probability density function $P(x_T, T; x_0, 0)$ for paths satisfying the δ-function constraint can be obtained by evaluating the expectation value $E(\delta(x(T) - x_T))$ as discussed in Section 3.2. This yields the result,

$$P(x_T, T; x_0, 0) = \left(2\pi \int_0^T f(T-t)^2 dt\right)^{-\frac{1}{2}}$$

$$\times \exp\left(-\left[\int_0^T f(T-t)^2 dt\right]^{-1} \frac{(x_T - x_0)^2}{2}\right). \quad (5.3)$$

There could be many possible choices for a memory function $f(T-t)$ and $h(t)$ that can be explored to model a given time series [25].

5.2 Mean Square Displacement with Memory

Experimental investigations often identify a relevant variable x and make many measurements of its mean square displacement as it changes in time thus, consequently, generating a time series. The mean square displacement (MSD), or variance [90], of a fluctuating variable x which measures deviations from a mean value $\langle x \rangle$ is given by (see Eq. (3.34) of Section 3.7),

$$\text{MSD} = \left\langle (x - \langle x \rangle)^2 \right\rangle$$

$$= \langle x^2 \rangle - \langle x \rangle^2. \quad (5.4)$$

As discussed in Section 3.7, we can use Eq. (5.3) to calculate the second moment expressed as,

$$\langle x^2 \rangle = \int_{-\infty}^{+\infty} x^2 \, P(x, T; x_0, 0) \, dx$$

$$= \left(2\pi \int_0^T [f(T-t) h(t)]^2 dt\right)^{-\frac{1}{2}}$$

$$\times \int_{-\infty}^{+\infty} x^2 \, \exp\left(-\left[\int_0^T [f(T-t) h(t)]^2 dt\right]^{-1} \frac{(x - x_0)^2}{2}\right) dx,$$

$$(5.5)$$

which gives,

$$\langle x^2 \rangle = x_0^2 + \int_0^T \left[f\left(T - t\right) h\left(t\right) \right]^2 dt\,. \tag{5.6}$$

With this, Eq. (5.4) becomes (let, $\langle x \rangle = x_0$),

$$\mathrm{MSD} = \int_0^T \left[f\left(T - t\right) h\left(t\right) \right]^2 dt\,. \tag{5.7}$$

We shall use this expression in discussing fluctuations in a time series.

5.2.1   *Wiener Process*

Clearly, when the memory function $f\left(T - t\right)$ is simply a constant $\sqrt{2D}$ and $h\left(t\right) = 1$, Eq. (5.7) yields the mean square displacement MSD_B of the Wiener process, or Brownian motion, i.e.,

$$\mathrm{MSD}_B = 2DT\,, \tag{5.8}$$

where D is the diffusion coefficient and T is time. A graph of MSD_B versus time is similar to the one shown in Figure 5.2 since the Wiener process is a special case of fractional Brownian motion.

5.2.2   *Fractional Brownian Motion*

To give another example, we now choose a memory function of the form,

$$f\left(T - t\right) = \frac{\left(T - t\right)^{H-1/2}}{\Gamma\left(H + 1/2\right)}\,, \tag{5.9}$$

with $h\left(t\right) = 1$. Equation (5.1) can then be written as,

$$x\left(T\right) = x_0 + B^H\left(T\right)\,, \tag{5.10}$$

where $B^H\left(T\right)$ is a fractional Brownian motion defined in the Riemann-Liouville representation by [156],

$$B^H\left(T\right) = \frac{1}{\Gamma\left(H + \frac{1}{2}\right)} \int_0^T \left(T - t\right)^{H-1/2} dB\left(t\right)\,. \tag{5.11}$$

In Eq. (5.11), H is the Hurst exponent with values $0 < H < 1$, as discussed in Section 3.4. With Eq. (5.9), the mean square displacement, Eq. (5.7), becomes,

$$\mathrm{MSD}_{fBm} = AT^\alpha\,, \tag{5.12}$$

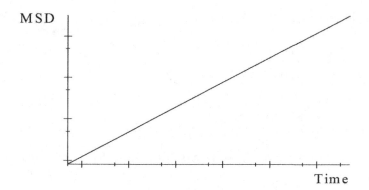

Fig. 5.2. A typical log-log plot of MSD versus time for fractional Brownian motion. The slope of the line is α.

where T is time, $\alpha = 2H$, and $A = 1/2H\ \Gamma\left(H + 1/2\right)^2$. For $H = \frac{1}{2}$, we recover ordinary Brownian motion.

A log-log plot of Eq. (5.12) for the Hurst exponent $H = 0.3$, which gives $\alpha = 0.6$ and $A = 1.2296$, is shown in Figure 5.2.

Equation (5.12) has been widely applied to explain processes exhibiting features of fractional Brownian motion.

5.2.3   *Exponentially-modified Brownian Motion*

Let us now consider a memory function $f\left(T - t\right)$ given by,

$$f\left(T - t\right) = \left(T - t\right)^{(\nu-1)/2},\qquad(5.13)$$

and $h\left(t\right)$ of the form,

$$h\left(t\right) = \frac{e^{-\mu t/2}}{t^{(1-\nu)/2}}.\qquad(5.14)$$

Using Eqs. (5.13) and (5.14) in Eq. (5.1), we have,

$$x\left(T\right) = x_0 + \int_0^T \frac{\left(T - t\right)^{(\nu-1)/2}e^{-\mu t/2}}{t^{(1-\nu)/2}}\ dB\left(t\right),\qquad(5.15)$$

which shows the Brownian motion $B\left(t\right)$ being exponentially modulated as time t ranges from 0 to T.

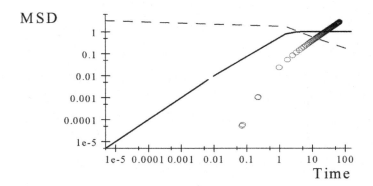

Fig. 5.3.   Log-log graph of MSD_{eB} versus time for $\mu = 1$, $\nu = 1$ (solid line); $\mu = 1$, $\nu = 1/2$ (dashed line); and $\mu = 5$, $\nu = 2$ (circle).

Multiplying Eqs. (5.13) and (5.14), and taking the square of the product, we get the integral,

$$
\int_0^T \left[f\left(T - t\right) h\left(t\right) \right]^2 dt = \int_0^T \frac{(T - t)^{\nu - 1} e^{-\mu t}}{t^{1 - \nu}} dt
$$

$$
= \frac{\sqrt{\pi}\,\Gamma\left(\nu\right) I_{\nu - \frac{1}{2}}\left(\frac{\mu t}{2}\right) e^{-(\mu t/2)}}{(t/\mu)^{\frac{1}{2} - \nu}}, \qquad (5.16)
$$

where we used Eq. (3.388.1) of [98], for $\mathrm{Re}\,\nu > 0$ and $T > 0$, $I_\lambda\left(z\right)$ is the modified Bessel function of the first kind, and $\Gamma\left(\mu\right)$ is the gamma function. Given Eq. (5.16), the mean square displacement Eq. (5.7), acquires the form,

$$
\mathrm{MSD}_{eB} = \frac{\sqrt{\pi}\,\Gamma\left(\nu\right) I_{\nu - \frac{1}{2}}\left(\frac{\mu t}{2}\right) e^{-(\mu t/2)}}{(t/\mu)^{\frac{1}{2} - \nu}}. \qquad (5.17)
$$

A log-log graph of MSD_{eB} versus t using Eq. (5.17) can be plotted for different value of μ and ν as shown in Figure 5.3.

The general behavior for large time exhibited by the graphs in Figure 5.3 may be understood using the asymptotic form of $I_\lambda\left(z\right)$ for large $|z|$ (Eq. (9.7.1) of [3]) given by $I_\lambda\left(z\right) \sim (2\pi z)^{-\frac{1}{2}} \exp\left[z - \left(\lambda^2 - \frac{1}{4}\right)/2z \right]$. This enables us to write, $\mathrm{MSD}_{eB} \sim \left(\Gamma\left(\nu\right)/\mu^\nu\right) t^{\nu - 1} \exp\left[-\nu\left(\nu - 1\right)/\mu t \right]$, the behavior of which becomes more transparent if we take its logarithm. We have,

$$
\ln\left(\mathrm{MSD}_{eB}\right) \sim \left(\nu - 1\right) \ln\left(t\right) + \ln\left(\frac{\Gamma\left(\nu\right)}{\mu^\nu}\right) - \frac{\nu\left(\nu - 1\right)}{\mu t}. \qquad (5.18)
$$

Designating, $y = \ln(\text{MSD}_{eB})$, $x = \ln(t)$, $m = \nu - 1$ and $b = \ln(\Gamma(\nu)/\mu^\nu)$, then Eq. (5.18) acquires the form,

$$y \sim mx + b - \left(\frac{\nu m}{\mu}\right) e^{-x}. \tag{5.19}$$

If we first ignore the last term on the right-hand side of Eq. (5.19), then the log-log graph of MSD_{eB} versus time is simply a straight line, i.e., $y = mx+b$, where the slope is $m = \nu - 1$. For several values of ν the slope of the line for large time, modified by the term $(\nu m/\mu)\, e^{-x}$, can be seen in Figure 5.3.

The shape of the graphs is in general retained if the time variable is rescaled from t to t/t_c, where t_c is a constant. This feature may help in modelling time series for a fluctuating physical variable. An example of log-log graphs with $t_c = 10$ is shown in Figure 5.4.

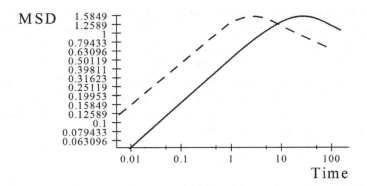

Fig. 5.4. Log-log graph for $\mu = 0.5$ and $\nu = 3/4$ (dashed line) and with rescaled time, t/t_c, where $t_c = 10$ (solid line).

We also note that from Eq. (5.19), where $m = \nu - 1$, varying the values of parameter ν changes the slope of the straight line segment of the graph as shown in Figure 5.5.

In modelling real-world phenomena, physical insights on the behavior of an exponentially modified Brownian motion as a stochastic process with memory would be helpful. In particular, the values of parameters μ, ν, and t_c in a model may be adjusted to match mean square displacements of a time series obtained from actual observations. We discuss in the next section an application to typhoon track fluctuations [30].

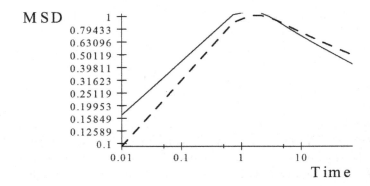

Fig. 5.5.   MSD versus time graph for $\mu = 1$ and $\nu = 0.75$ (solid line); $\nu = 0.8$ (dashed line).

5.3   Typhoon Track Fluctuations

Accurate predictions of tropical typhoon trajectories remain to be a challenging task in spite of the application of advanced numerical simulations [54; 89] and various statistical analyses [47; 101]. In this section, we see how the white noise functional integral with a suitable memory function can give an analytical description of the translational motion of a typhoon. We first decompose the typhoon position in time into two parts: a kinematic deterministic component and a random fluctuating part. Thus, the location $\mathbf{r}$ of a typhoon is estimated using the equation,

$$\mathbf{r} = \mathbf{v}t + \frac{1}{2}\mathbf{a}t^2 + \Delta\mathbf{r}\,, \qquad (5.20)$$

where $\mathbf{r} = (x, y)$, marks points along the longitude and latitude, and $\Delta\mathbf{r}$ is fluctuation arising from ambient temperature gradients in the atmosphere, as well as wind movements from surrounding pockets of cold and hot winds, among other factors. The first two terms on the right-hand side of Eq. (5.20) comprise the deterministic part and t is elapsed time where $\mathbf{v}$ is velocity determined from the typhoon's immediate history. The acceleration $\mathbf{a}$ would essentially result from Corioli's force which depends on the earth's rotation and velocity $\mathbf{v}$ of a typhoon.

The stochastic component $\Delta\mathbf{r} = (\Delta x, \Delta y)$ in Eq. (5.20) has been a subject of several investigations [30; 154]. It has been noted that fluctuations of typhoon tracks in a given geographical area seem to follow patterns that appear to obey a universal law [215; 154]. In [154] for example, the longitude and latitude of cyclones, also referred to as hurricanes or typhoons, were plotted against time to analyze fluctuations from the cyclone's mean track. Designating the position of the cyclone as x, [154] essentially investigated a power law corresponding to the mean square deviation (MSD), $\left\langle [x(t+t') - x(t)]^2 \right\rangle \sim t^\alpha$, where t is time and the exponent α takes values smaller (subdiffusive) or larger (superdiffusive) than 1. Note that this power law is essentially Eq. (5.12) for fractional Brownian motion. The results in [154] exhibited a considerable spread in the values of α, and a histogram showed a peak-value of $\alpha = 1.65$. In a log-log plot of MSD versus time, however, a power law similar to Eq. (5.12) is not able to account for a downward curve appearing at longer times for almost all cyclones investigated. In [154], the authors attributed this curve to a probable lack of statistics. This motivates us to take another type of memory function $f(T-t)$ in combination with $h(t)$ which could account for downward curves appearing at longer times in MSD versus time graphs observed for cyclone tracks [30].

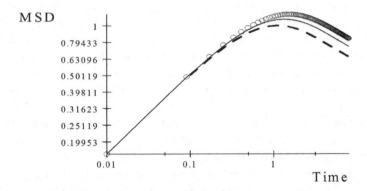

Fig. 5.6. Mean square displacement for an exponentially-modified Brownian motion with $\nu = 3/4$, and $\mu = 1.2$ (dashed line); $\mu = 1$ (solid line); and $\mu = 0.87$ (circle).

As discussed in Section 5.2.3, downward curves appearing at longer times in MSD versus time graphs for cyclone tracks can be described by an exponentially-modified Brownian motion. Figure 5.6, for example, shows a downward curve at longer times which may give a better fit for the curves of the nine cyclones shown in Figures 3a and 3b of reference [154]. This suggests that the memory behavior of fluctuations in cyclone tracks can only be approximately described by a power law associated with fractional Brownian motion, i.e., $\left\langle [x(t+t') - x(t)]^2 \right\rangle \sim t^\alpha$, which was adopted by

T. Meuel *et al.* [154]. Fluctuations described by an exponentially modified Brownian motion on the other hand, as shown by Eq. (5.15), may instead help us better understand cyclone track fluctuations and forecast cones [30].

5.4 Particle-tracking in Microrheology

Recent techniques to measure rheological properties of complex fluids such as viscosity are often done by tracking the trajectory of a micrometer size particle suspended in the medium. The medium could be a colloidal suspension, a soft biological sample such as the cytoplasm of a living cell, polymer solution and gel, among many others. By studying the motion of the particle when the medium is subjected to stress, insights into the structural and mechanical response of various materials are obtained. For instance, thermally driven random motion of particles in a complex fluid differs from that of diffusive Brownian motion of similar particles suspended in a purely viscous fluid [150; 151; 200]. A tracer particle which freely diffuses normally indicates that the medium is purely viscous, while a subdiffusive path evolution indicates viscoelastic properties. Experimental verification of these involves microrheological measurements of high spatial and temporal resolution that can record time-dependent trajectories of a microparticle suspended in fluids [200].

The measurements done in experiments could be in the form of passive microrheology, which tracks movement of the probe particle brought about by thermal fluctuations of the medium's molecules, or active rheology where external forces are also applied to the particle using, for example, optical tweezers [114; 200]. When the applied stress is removed, a viscoelastic material returns to its original form as a function of elapsed time. Such return after deformation, in fact, signals memory properties of viscoelastic materials. This motivates us to present an analytical model dealing with diffusive paths of probe particles in a medium exhibiting memory.

In various microrheological experiments, the MSD $\left\langle (\Delta r)^2 \right\rangle$ of the particle versus time is measured. From this, the viscoelastic moduli, for example, are obtained using the generalized Stokes-Einstein relation [150]. Fluctuations of probe particles in fluids, however, seem to follow a common pattern as shown in experimental plots of MSD versus time. Similarities, for example, are exhibited in Figure 1 of [150] for concentrated monodisperse emulsion droplets of radius $a = 0.53\mu$m, and Figure 4 of [200] for an optically trapped 4.74μm diameter silica bead suspended in water. Figure 1 (inset) of [151] for hard spheres in complex fluids also shows a similarly curved graph of MSD versus time. In this section, we model these types of time-dependent MSD as path fluctuations of a probe particle with memory properties. In particular, we employ an exponentially modified Brownian motion to explain this common pattern appearing in particle tracking experiments.

Let us designate the position of a probe particle in a fluid as, $r(t) = (x(t), y(t), z(t))$. Without loss of generality, we simply consider the x-coordinate of the probe particle which we parametrize as Eq. (5.1) using a memory function $f(T-t)$ to record the drag as t varies from 0 to T. As shown in Section 5.1, the probability density function and the mean square displacement can be obtained from Eqs. (5.3) and (5.7), respectively.

We now consider an exponentially modified Brownian motion which appears appropriate for describing the observed curves for log-log plots of MSD versus time of probe particles in complex fluids [150; 151; 200].

5.4.1 *Exponentially-modified Brownian Motion in Viscoelastic Materials*

Let us now consider the memory function $f(T-t)$ and $h(t)$ given by,

$$f(T-t)h(t) = (T-t)^{(\mu-1)/2} e^{-\beta/2t} t^{-\mu}. \qquad (5.21)$$

Using this combination in Eq. (5.1), we have,

$$x(T) = x_0 + \int_0^T (T-t)^{(\mu-1)/2} e^{-\beta/2t} t^{-\mu} \, dB(t), \qquad (5.22)$$

which shows Brownian motion $B(t)$ exponentially modified as time t ranges from 0 to T. Note, however, that this exponentially modified Brownian motion is slightly different from Eq. (5.15).

The integral of the square of Eq. (5.21) is given by,

$$\int_0^T [f(T-t)h(t)]^2 \, dt = \int_0^T (T-t)^{\mu-1} e^{-\beta/t} t^{-2\mu} \, dt$$

$$= \frac{1}{\sqrt{\pi T}} \beta^{\frac{1}{2}-\mu} e^{-\frac{\beta}{2T}} \Gamma(\mu) K_{\mu-\frac{1}{2}} \left(\frac{\beta}{2T}\right), \qquad (5.23)$$

where we used Eq. (3.471.4) of [98], for Re$\mu > 0$ and $T > 0$. In Eq. (5.23), $\Gamma(\mu)$ is the gamma function and $K_\nu(z)$ is the modified Bessel function of the second kind. Using Eq. (5.23) in Eq. (5.3), the probability density function corresponding to this exponentially modified Brownian motion can then be obtained [25].

With Eq. (5.23), the mean square displacement Eq. (5.7), acquires the form,

$$\text{MSD} = \frac{1}{\sqrt{\pi T}} \beta^{\frac{1}{2}-\mu} e^{-\frac{\beta}{2T}} \Gamma\left(\mu\right) K_{\mu-\frac{1}{2}}\left(\frac{\beta}{2T}\right). \qquad (5.24)$$

5.4.2   *Mean Square Displacement from Microrheological Experiments*

A log-log graph of MSD versus time t using Eq. (5.24) can be plotted for $\mu = \frac{1}{2}$ and $\beta = 50$, as shown in Figure 5.7.

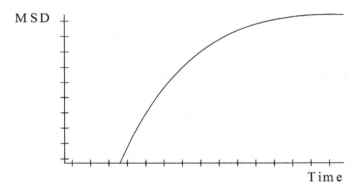

Fig. 5.7.   Log-log plot of MSD versus time for $\mu = 1/2$ and $\beta = 50$.

The graph given by Figure 5.7 shows the general shape of MSD versus time graphs of particle-tracking data from microrheological experiments. This is exemplified by data on emulsion droplets of radius $a = 0.53\mu$m (Figure 1 of [150]), as well as for an optically trapped 4.74μm diameter silica bead in water (Figure 4, [200]), and likewise Figure 1 (inset) of [151]. Fine tuning of the graph for specific cases may be done by varying the values of μ and β in Eq. (5.24). Apparently, the behavior of micro particle fluctuations in complex fluids may be described by an exponentially modified Brownian motion, i.e., Eq. (5.22), which could provide additional insight into the nature of viscoelastic materials. Other memory functions discussed in Chapter 3 can also be explored.

Chapter 6

Fluctuations without Memory

Many examples abound in nature which exhibit fluctuations with apparent no memory of the past. A classic example of this is Brownian motion which is observed when a small particle is immersed in a fluid. The position of the particle fluctuates since molecules of the fluid frequently collide with the particle in an unpredictable way. In many problems which involve random fluctuations, the Fokker-Planck equation has been a powerful tool in investigating many natural phenomena [181]. The first application of the Fokker-Planck equation was, in fact, the Brownian motion where a probability density for the position of a particle can be obtained [88; 174].

In the context of earlier discussions, we define fluctuations without memory as a case where the memory function is, $f(\tau - t) = 1$ and $h(t) = 1$. In particular, Eq. (1.12) given by, $x(\tau) = x_0 + \int f(\tau - t)h(t)\omega(t)dt$, reduces to that of Eq. (1.16) where fluctuations are described by a Brownian motion, $B(t) = \int \omega(\tau)d\tau$, with $\omega(\tau)$ a Gaussian white noise variable. We shall illustrate in this chapter that white noise calculus, combined with Feynman's path integral, allows one to arrive at solutions to the Fokker-Planck equation in a relatively direct manner. Before proceeding to the Fokker-Planck equation, however, we first look at its relation to a stochastic differential equation.

6.1 Correspondence between a Stochastic Differential Equation and the Fokker-Planck Equation

Let us consider a random process $X(t)$ which satisfies Ito's stochastic differential equation of the form[1],

$$dX(t) = a(t, X(t)) dt + b(t, X(t)) dB(t), \qquad (6.1)$$

[1]The authors thank Prof. Hiroshi Ezawa of Gakushuin University for his permission to base this section on a seminar he gave at the National Institute of Physics, University of the Philippines.

where $B(t)$ is the Wiener process, t is time ($0 \leq t \leq +\infty$). The solution to Eq. (6.1) is a diffusion process where the coefficient of drift is $a(t, X(t))$, and the coefficient of diffusion is, $[b(t, X(t))]^2$.

We shall use the fact that the solution to Eq. (6.1) is a Markov process, that is, its future behavior depends only on information about its present state and not on its past state. We can then associate to $X(t)$ a probability density function $P(x, t; x_0, t_0)$ which satisfies the Chapman-Kolmogorov condition,

$$P(x, t; x_0, t_0) = \int_{-\infty}^{+\infty} P(x, t; x', t') P(x', t'; x_0, t_0) \, dx'. \qquad (6.2)$$

With this, it shall be shown that $P(x, t; x_0, t_0)$ satisfies the Fokker-Planck equation.

Let us take an arbitrary continuous bounded function, $\varphi(x)$, with continuous bounded first and second derivatives such that at $|x| = \infty$, we have, $\varphi(x) = 0$, and $d\varphi(x)/dx = 0$. The conditional expectation of $\varphi(x')$ at time, $t' = t + dt$, can be written as,

$$\langle \varphi(x') \mid x(t_0) = x_0 \rangle = \int_{-\infty}^{+\infty} \varphi(x') P(x', t'; x_0, t_0) \, dx'. \qquad (6.3)$$

Applying the Chapman-Kolmogorov condition, Eq. (6.2), to rewrite $P(x', t'; x_0, t_0)$ in Eq. (6.3), we get,

$$\langle \varphi(x') \mid x(t_0) = x_0 \rangle = \int_{-\infty}^{+\infty} \varphi(x') \left[\int_{-\infty}^{+\infty} P(x', t'; x, t) \; P(x, t; x_0, t_0) \, dx \right] dx'$$

$$= \int_{-\infty}^{+\infty} \left\{ \int_{-\infty}^{+\infty} \varphi(x') P(x', t'; x, t) \, dx' \right\} P(x, t; x_0, t_0) \, dx. \qquad (6.4)$$

The term in the curly brackets is just the expectation of $\varphi(x')$, and Eq. (6.4) acquires the form,

$$\langle \varphi(x') \mid x(t_0) = x_0 \rangle = \int_{-\infty}^{+\infty} \langle \varphi(x') \mid x(t) = x \rangle P(x, t; x_0, t_0) \, dx. \qquad (6.5)$$

Consider $\varphi(x')$ on the right-hand side of Eq. (6.5). An expansion of $\varphi(x')$ about x in a Taylor series yields,

$$\varphi(x') = \varphi(x')|_x + \frac{d\varphi(x')}{dx}\bigg|_x [(x+dx)-x]$$

$$+\frac{1}{2}\frac{d^2\varphi(x')}{dx^2}\bigg|_x [(x+dx)-x]^2 + \dots . \qquad (6.6)$$

Keeping only up to second-order terms we get,

$$\varphi(x') = \varphi(x) + \varphi'(x)\,dx + \frac{1}{2}\varphi''(x)\,(dx)^2, \qquad (6.7)$$

with $\varphi'(x) = d\varphi(x)/dx$. Note, however, that dx is given by Eq. (6.1), i.e.,

$$dx = a\,dt + b\,d, \qquad (6.8)$$

where, for simplicity of notation, we write a and b instead of $a(t, X(t))$ and $b(t, X(t))$, respectively. With Eq. (6.8), we can write Eq. (6.7) as,

$$\varphi(x') = \varphi(x) + \varphi'(x)\,(a\,dt + b\,dB)$$

$$+\frac{1}{2}\varphi''(x)\,(a\,dt + b\,dB)^2. \qquad (6.9)$$

Neglecting the term of second-order in (dt), Eq. (6.9) becomes,

$$\varphi(x') = \varphi(x) + \varphi'(x)\,a\,dt + \varphi'(x)\,b\,dB$$

$$+\frac{1}{2}\varphi''(x)\,b^2\,(dB)^2 + \varphi''(x)\,ab\,dt\,dB. \qquad (6.10)$$

We now take the ensemble average of Eq. (6.10) and use the properties of the standard Wiener process, i.e., $\langle dB \rangle = 0$ and $\left\langle (dB)^2 \right\rangle = dt$, to obtain,

$$\langle \varphi(x') \rangle = \varphi(x) + \varphi'(x)\,a\,dt + \frac{1}{2}\varphi''(x)\,b^2 dt. \qquad (6.11)$$

Using Eq. (6.11) for $\langle \varphi (x') \mid x (t) = x \rangle$ on the right-hand side of Eq. (6.5), we have,

$$\langle \varphi (x') \mid x (t_0) = x_0 \rangle = \int\limits_{-\infty}^{+\infty} \varphi (x) \, P (x, t; x_0, t_0) \, dx$$

$$+ \int\limits_{-\infty}^{+\infty} \left\{ \frac{d\varphi (x)}{dx} a + \frac{1}{2} \frac{d^2\varphi (x)}{dx^2} b^2 \right\} P (x, t; x_0, t_0) \, dt \, dx \, .$$

$$(6.12)$$

We observe that the left-hand side of this equation can also be written as,

$$\langle \varphi (x') \mid x (t_0) = x_0 \rangle = \int\limits_{-\infty}^{+\infty} \varphi (x) \, P (x, t + dt; x_0, t_0) \, dx \, . \qquad (6.13)$$

Substituting Eq. (6.13) for the left-hand side of Eq. (6.12), we get,

$$\int\limits_{-\infty}^{+\infty} \varphi (x) \, P (x, t + dt; x_0, t_0) \, dx = \int\limits_{-\infty}^{+\infty} \varphi (x) \; P (x, t; x_0, t_0) \, dx$$

$$+ \int\limits_{-\infty}^{+\infty} \left\{ \frac{d\varphi (x)}{dx} a + \frac{1}{2} \frac{d^2\varphi (x)}{dx^2} b^2 \right\}$$

$$\times P (x, t; x_0, t_0) \, dt \, dx \, . \qquad (6.14)$$

Transferring the first term on the right to the left-hand side of the equation, and dividing by dt gives,

$$\int\limits_{-\infty}^{+\infty} \varphi (x) \left\{ \frac{P (x, t + dt; x_0, t_0) - P (x, t; x_0, t_0)}{dt} \right\} dx$$

$$= \int\limits_{-\infty}^{+\infty} \left\{ \frac{d\varphi (x)}{dx} a + \frac{1}{2} \frac{d^2\varphi (x)}{dx^2} b^2 \right\} P (x, t; x_0, t_0) \, dx \, . \qquad (6.15)$$

This can further be expressed as,

$$\int\limits_{-\infty}^{+\infty} \varphi (x) \frac{\partial P (x, t; x_0, t_0)}{\partial t} dx = \int\limits_{-\infty}^{+\infty} \left\{ \frac{d\varphi (x)}{dx} a + \frac{1}{2} \frac{d^2\varphi (x)}{dx^2} b^2 \right\}$$

$$\times P (x, t; x_0, t_0) \, dx \, . \qquad (6.16)$$

Consider the first integral on the right-hand side of Eq. (6.16) which can be integrated by parts as,

$$\int\limits_{-\infty}^{+\infty} \left[\frac{d\varphi(x)}{dx}a\right] P(x,t;x_0,t_0)\,dx = \int\limits_{-\infty}^{+\infty} \frac{d}{dx}\left[\varphi(x)\,aP(x,t;x_0,t_0)\right]dx$$

$$- \int\limits_{-\infty}^{+\infty} \varphi(x)\frac{\partial}{\partial x}\left[aP(x,t;x_0,t_0)\right]\,dx$$

$$= \left[\varphi(x)\,aP(x,t;x_0,t_0)\right]|_{-\infty}^{+\infty}$$

$$- \int\limits_{-\infty}^{+\infty} \varphi(x)\frac{\partial}{\partial x}\left[aP(x,t;x_0,t_0)\right]dx\,.$$

$$(6.17)$$

Using the fact that $\varphi(x) = 0$ at $|x| = \infty$, the first term on the right gives zero and Eq. (6.17) becomes,

$$\int\limits_{-\infty}^{+\infty} \left[\frac{d\varphi(x)}{dx}a\right] P(x,t;x_0,t_0)\,dx = -\int\limits_{-\infty}^{+\infty} \varphi(x)\frac{\partial}{\partial x}\left[aP(x,t;x_0,t_0)\right]dx\,.$$

$$(6.18)$$

Likewise, the second term on the right-hand side of Eq. (6.16) can be integrated by parts to yield,

$$\frac{1}{2}\int\limits_{-\infty}^{+\infty} \frac{d^2\varphi(x)}{dx^2}b^2 P(x,t;x_0,t_0)\,dx = \frac{1}{2}\int\limits_{-\infty}^{+\infty} \frac{d}{dx}\left[\frac{d\varphi(x)}{dx}b^2 P(x,t;x_0,t_0)\right]dx$$

$$-\frac{1}{2}\int\limits_{-\infty}^{+\infty} \frac{d\varphi(x)}{dx}\frac{d}{dx}\left[b^2 P(x,t;x_0,t_0)\right]dx$$

$$= \frac{1}{2}\left[\frac{d\varphi(x)}{dx}b^2 P(x,t;x_0,t_0)\right]\Big|_{-\infty}^{+\infty}$$

$$-\frac{1}{2}\int\limits_{-\infty}^{+\infty} \frac{d\varphi(x)}{dx}\frac{d}{dx}\left[b^2 P(x,t;x_0,t_0)\right]dx$$

$$= 0 - \frac{1}{2}\int\limits_{-\infty}^{+\infty} \frac{d\varphi(x)}{dx}\frac{d}{dx}\left[b^2 P(x,t;x_0,t_0)\right]dx\,,$$

$$(6.19)$$

since $d\varphi(x)/dx = 0$ at $|x| = \infty$. The remaining integral on the right-hand side can again be evaluated using integration by parts and we have,

$$\int\limits_{-\infty}^{+\infty} \frac{d^2\varphi(x)}{dx^2} b^2 P(x,t;x_0,t_0)\, dx = -\int\limits_{-\infty}^{+\infty} \frac{d}{dx}\left[\varphi(x)\frac{d}{dx}\left(b^2 P(x,t;x_0,t_0)\right)\right] dx$$

$$+ \int\limits_{-\infty}^{+\infty} \varphi(x)\frac{d^2}{dx^2}\left[b^2 P(x,t;x_0,t_0)\right] dx$$

$$= -\left\{\varphi(x)\frac{d}{dx}\left[b^2 P(x,t;x_0,t_0)\right]\right\}\Bigg|_{-\infty}^{+\infty}$$

$$+ \int\limits_{-\infty}^{+\infty} \varphi(x)\frac{d^2}{dx^2}\left[b^2 P(x,t;x_0,t_0)\right] dx\,. \tag{6.20}$$

Knowing that, $\varphi(x) = 0$ at $|x| = \infty$, the term with curly brackets gives zero, and Eq. (6.20) reduces to,

$$\frac{1}{2}\int\limits_{-\infty}^{+\infty} \frac{d^2\varphi(x)}{dx^2} b^2 P(x,t;x_0,t_0)\, dx = \frac{1}{2}\int\limits_{-\infty}^{+\infty} \varphi(x)\frac{d^2}{dx^2}\left[b^2 P(x,t;x_0,t_0)\right] dx\,. \tag{6.21}$$

Combining the results Eqs. (6.18) and (6.21), we rewrite the right-hand side of Eq. (6.16) to get the expression,

$$\int\limits_{-\infty}^{+\infty} \varphi(x)\frac{\partial P(x,t;x_0,t_0)}{\partial t}\, dx = -\int\limits_{-\infty}^{+\infty} \varphi(x)\frac{\partial}{\partial x}\left[aP(x,t;x_0,t_0)\right] dx$$

$$+ \frac{1}{2}\int\limits_{-\infty}^{+\infty} \varphi(x)\frac{d^2}{dx^2}\left[b^2 P(x,t;x_0,t_0)\right] dx\,. \tag{6.22}$$

Putting everything on one side of the equation we arrive at,

$$0 = \int\limits_{-\infty}^{+\infty} \varphi(x)\left\{\frac{\partial P(x,t;x_0,t_0)}{\partial t} + \frac{\partial}{\partial x}\left[aP(x,t;x_0,t_0)\right]\right.$$

$$\left. - \frac{1}{2}\frac{d^2}{dx^2}\left[b^2 P(x,t;x_0,t_0)\right]\right\} dx\,. \tag{6.23}$$

Since $\varphi(x)$ is arbitrary, Eq. (6.23) tells us that,

$$\left\{\frac{\partial P(x,t;x_0,t_0)}{\partial t} + \frac{\partial}{\partial x}\left[aP(x,t;x_0,t_0)\right] - \frac{1}{2}\frac{d^2}{dx^2}\left[b^2 P(x,t;x_0,t_0)\right]\right\} = 0.$$

(6.24)

Equation (6.24) is the forward Fokker-Planck equation for the probability density function $P(x,t;x_0,t_0)$ associated with the stochastic process $X(t)$ of Eq. (6.1). In this equation, the drift coefficient is a, and b^2 is the diffusion coefficient.

6.2 Short-Time Solution for the Fokker-Planck Equation

We now look at the short-time solution to the Fokker-Planck equation which leads to long-time solutions expressed as a path integral[2]. Let us first derive the short-time solution of a d-dimensional Fokker-Planck equation. Writing the components of the drift vector $\mathbf{A}(\mathbf{x},t)$ as $a_k(\mathbf{x},t)$, and the elements of diffusion matrix $\mathbb{B}(\mathbf{x},t)$ as $b_{kl}(\mathbf{x},t)$, where $k,l = 1,2,...,d$, and $\mathbf{x} = (x_1,x_2,...,x_d)$, the Fokker-Planck equation can be written as,

$$\partial_t P(\mathbf{x},t \mid \mathbf{x}_0,t_0) = -\partial_k\left[a_k(\mathbf{x},t) P(\mathbf{x},t \mid \mathbf{x}_0,t_0)\right]$$

$$+\frac{1}{2}\partial_{k,l}\left[b_{kl}(\mathbf{x},t) P(\mathbf{x},t \mid \mathbf{x}_0,t_0)\right]. \quad (6.25)$$

Here, we employ the notation, $\partial_t = \partial/\partial t$, $\partial_k = \partial/\partial x_k$, $\partial_{k,l} = (\partial/\partial x_k)(\partial/\partial x_l)$, and summation over repeated indices is assumed. In Eq. (6.25), the conditional probability density $P(\mathbf{x},t \mid \mathbf{x}_0,t_0)$ obeys the initial condition,

$$\lim_{t \to t_0} P(\mathbf{x},t \mid \mathbf{x}_0,t_0) = \delta(\mathbf{x} - \mathbf{x}_0). \quad (6.26a)$$

Let us now consider short intervals of time for the conditional probability, $P(\mathbf{x},t+\Delta t \mid \mathbf{x}_0,t)$, where Δt is infinitesimal and write the expansion,

$$P(\mathbf{x},t+\Delta t \mid \mathbf{x}_0,t) = P(\mathbf{x},t \mid \mathbf{x}_0,t) + \left[\partial_t P(\mathbf{x},t \mid \mathbf{x}_0,t)\right]\Delta t + O\left((\Delta t)^2\right).$$

(6.27)

---

[2]This section is based on derivations made by Matthew G. O. Escobido during a visit at the Research Center for Theoretical Physics, Central Visayan Institute Foundation. The authors wish to thank him for his permission to reproduce his discussions on the topic.

For the second term on the right-hand side, the factor $\partial_t P\left(\mathbf{x}, t \mid \mathbf{x}_0, t\right)$ can be expressed using Eq. (6.25). Moreover, wherever there is $P\left(\mathbf{x}, t \mid \mathbf{x}_0, t\right)$, we replace it with $\delta\left(\mathbf{x} - \mathbf{x}_0\right)$ in view of the initial condition Eq. (6.26a). We then write Eq. (6.27) as,

$$P\left(\mathbf{x}, t + \Delta t \mid \mathbf{x}_0, t\right) \cong \delta\left(\mathbf{x} - \mathbf{x}_0\right) - \partial_k\left[a_k\left(\mathbf{x}, t\right) \delta\left(\mathbf{x} - \mathbf{x}_0\right)\right] \Delta t$$

$$+ \frac{1}{2} \partial_{k,l}\left[b_{kl}\left(\mathbf{x}, t\right) \delta\left(\mathbf{x} - \mathbf{x}_0\right)\right] \Delta t, \tag{6.28}$$

where we dropped terms of order $(\Delta t)^2$ and higher. We now take the Fourier transform of Eq. (6.28), i.e.,

$$\mathcal{F}\left[P\left(\mathbf{x}, t + \Delta t \mid \mathbf{x}_0, t\right)\right] = \left(\frac{1}{2\pi}\right)^{d/2} \int\limits_{-\infty}^{+\infty} \exp\left(i\mathbf{w} \cdot \mathbf{x}\right) P\left(\mathbf{x}, t + \Delta t \mid \mathbf{x}_0, t\right) d\mathbf{x}$$

$$= I_1 - I_2 + I_3, \tag{6.29}$$

where we use the right-hand side of Eq. (6.28) to define the integrals,

$$I_1 = \left(\frac{1}{2\pi}\right)^{d/2} \int\limits_{-\infty}^{+\infty} \exp\left(i\mathbf{w} \cdot \mathbf{x}\right) \delta\left(\mathbf{x} - \mathbf{x}_0\right) d\mathbf{x}$$

$$= \left(\frac{1}{2\pi}\right)^{d/2} \exp\left(i\mathbf{w} \cdot \mathbf{x}_0\right), \tag{6.30}$$

where evaluation of I_1 is facilitated by the delta function, and

$$I_2 = \left(\frac{1}{2\pi}\right)^{d/2} \int\limits_{-\infty}^{+\infty} \exp\left(i\mathbf{w} \cdot \mathbf{x}\right) \partial_k\left[a_k\left(\mathbf{x}, t\right) \delta\left(\mathbf{x} - \mathbf{x}_0\right)\right] d\mathbf{x} \Delta t; \tag{6.31}$$

$$I_3 = \left(\frac{1}{2\pi}\right)^{d/2} \frac{1}{2} \int\limits_{-\infty}^{+\infty} \exp\left(i\mathbf{w} \cdot \mathbf{x}\right) \partial_{k,l}\left[b_{kl}\left(\mathbf{x}, t\right) \delta\left(\mathbf{x} - \mathbf{x}_0\right)\right] \Delta t \, d\mathbf{x}. \tag{6.32}$$

Let us first consider I_2. We rewrite Eq. (6.31) as,

$$I_2 = \left(\frac{1}{2\pi}\right)^{d/2} \int\limits_{-\infty}^{+\infty} \partial_k\left\{\exp\left(i\mathbf{w} \cdot \mathbf{x}\right) a_k\left(\mathbf{x}, t\right) \delta\left(\mathbf{x} - \mathbf{x}_0\right)\right\} d\mathbf{x} \Delta t$$

$$- \left(\frac{1}{2\pi}\right)^{d/2} \int\limits_{-\infty}^{+\infty} a_k\left(\mathbf{x}, t\right) \delta\left(\mathbf{x} - \mathbf{x}_0\right) \partial_k\left[\exp\left(i\mathbf{w} \cdot \mathbf{x}\right)\right] d\mathbf{x} \Delta t. \tag{6.33}$$

The first integral in Eq. (6.33) turns out to be zero. This can be shown as follows. With $\mathbf{w} \cdot \mathbf{x} = w_j x_j = w_1 x_1 + w_2 x_2 + \ldots + w_d x_d$, we can run the repeated index k and, for clarity, write the first term of Eq. (6.33) explicitly as,

$$\int\limits_{-\infty}^{+\infty} \partial_k \left\{ \exp\left(i\mathbf{w} \cdot \mathbf{x} \right) a_k \left(\mathbf{x}, t \right) \delta \left(\mathbf{x} - \mathbf{x}_0 \right) \right\} d\mathbf{x} \Delta t$$

$$= \int\limits_{-\infty}^{+\infty} \cdots \int \frac{\partial}{\partial x_1} \left[e^{w_1 x_1 + w_2 x_2 + \ldots + w_d x_d} a_1 \left(\mathbf{x}, t \right) \delta \left(\mathbf{x} - \mathbf{x}_0 \right) \right] dx_1 dx_2 \cdots dx_d \Delta t$$

$$+ \int\limits_{-\infty}^{+\infty} \cdots \int \frac{\partial}{\partial x_2} \left[e^{w_1 x_1 + w_2 x_2 + \ldots + w_d x_d} a_2 \left(\mathbf{x}, t \right) \delta \left(\mathbf{x} - \mathbf{x}_0 \right) \right] dx_1 dx_2 \cdots dx_d \Delta t$$

$$+ \quad \cdots$$

$$+ \int\limits_{-\infty}^{+\infty} \cdots \int \frac{\partial}{\partial x_d} \left[e^{w_1 x_1 + w_2 x_2 + \ldots + w_d x_d} a_d \left(\mathbf{x}, t \right) \delta \left(\mathbf{x} - \mathbf{x}_0 \right) \right] dx_1 dx_2 \cdots dx_d \Delta t .$$

$$(6.34)$$

Each term in Eq. (6.34) when evaluated gives zero. Consider for example the second term where the integral over x_2 can immediately be calculated, i.e.,

$$\int\limits_{-\infty}^{+\infty} \cdots \int \frac{\partial}{\partial x_2} \left[e^{w_1 x_1 + w_2 x_2 + \ldots + w_d x_d} a_2 \left(\mathbf{x}, t \right) \delta \left(\mathbf{x} - \mathbf{x}_0 \right) \right] dx_1 dx_2 \cdots dx_d \Delta t$$

$$= \int\limits_{-\infty}^{+\infty} \cdots \int \left[e^{w_1 x_1 + w_2 x_2 + \ldots + w_d x_d} a_1 \left(\mathbf{x}, t \right) \delta \left(\mathbf{x} - \mathbf{x}_0 \right) \right]_{x_2 = -\infty}^{x_2 = +\infty} dx_1 dx_3 \cdots dx_d \Delta t .$$

$$(6.35)$$

Since $\delta \left(\mathbf{x} - \mathbf{x}_0 \right) = \delta \left(x_1 - (x_1)_0 \right) \delta \left(x_2 - (x_2)_0 \right) \cdots \delta \left(x_d - (x_d)_0 \right)$, the value of the delta function $\delta \left(x_2 - (x_2)_0 \right)$ is zero when evaluated at $x_2 = \pm\infty$ since, physically, the initial point $(x_2)_0$ is not at infinity, hence, $x_2 \neq (x_2)_0$. A similar situation occurs giving zero for each term in Eq. (6.34) and, therefore, Eq. (6.33) consists simply of the second term which becomes,

$$I_2 = -\left(\frac{1}{2\pi}\right)^{d/2} \int\limits_{-\infty}^{+\infty} a_k\left(\mathbf{x}, t\right) \delta\left(\mathbf{x} - \mathbf{x}_0\right) \partial_k \left[\exp\left(i\mathbf{w} \cdot \mathbf{x}\right)\right] d\mathbf{x}\Delta t$$

$$= -\left(\frac{1}{2\pi}\right)^{d/2} \int\limits_{-\infty}^{+\infty} a_k\left(\mathbf{x}, t\right) \delta\left(\mathbf{x} - \mathbf{x}_0\right) \left[\exp\left(i\mathbf{w} \cdot \mathbf{x}\right)\right] \left(iw_k\right) d\mathbf{x}\Delta t$$

$$= -\left(\frac{1}{2\pi}\right)^{d/2} a_k\left(\mathbf{x}_0, t\right) \left[\exp\left(i\mathbf{w} \cdot \mathbf{x}_0\right)\right] \left(iw_k\right) \Delta t$$

$$= -i \left(\frac{1}{2\pi}\right)^{d/2} \mathbf{w} \cdot \mathbf{A}\left(\mathbf{x}_0, t\right) \left[\exp\left(i\mathbf{w} \cdot \mathbf{x}_0\right)\right] \Delta t . \tag{6.36}$$

We now proceed and consider the last integral I_3, Eq. (6.32), where the integral can be written as,

$$\int\limits_{-\infty}^{+\infty} \exp\left(i\mathbf{w} \cdot \mathbf{x}\right) \partial_{k,l} \left[b_{kl}\left(\mathbf{x}, t\right) \delta\left(\mathbf{x} - \mathbf{x}_0\right)\right] d\mathbf{x}\Delta t$$

$$= \int\limits_{-\infty}^{+\infty} \partial_k \left\{\exp\left(i\mathbf{w} \cdot \mathbf{x}\right) \partial_l \left[b_{kl}\left(\mathbf{x}, t\right) \delta\left(\mathbf{x} - \mathbf{x}_0\right)\right]\right\} d\mathbf{x}\Delta t$$

$$- \int\limits_{-\infty}^{+\infty} \left[\partial_k \exp\left(i\mathbf{w} \cdot \mathbf{x}\right)\right] \partial_l \left[b_{kl}\left(\mathbf{x}, t\right) \delta\left(\mathbf{x} - \mathbf{x}_0\right)\right] d\mathbf{x}\Delta t . \tag{6.37}$$

The first integral of Eq. (6.37) again turns out to be zero, since for any value of $k = 1, 2, ..., d$, (as in Eq. (6.34)) we have,

$$\int\limits_{-\infty}^{+\infty} \frac{\partial}{\partial x_k} \left\{\exp\left(i\mathbf{w} \cdot \mathbf{x}\right) \frac{\partial}{\partial x_l} \left(b_{kl}\left(\mathbf{x}, t\right) \delta\left(\mathbf{x} - \mathbf{x}_0\right)\right)\right\} dx_1 dx_2...dx_k...dx_d \Delta t$$

$$= \int\limits_{-\infty}^{+\infty} \left[\exp\left(i\mathbf{w} \cdot \mathbf{x}\right) \frac{\partial}{\partial x_l} \left(b_{kl}\left(\mathbf{x}, t\right) \delta\left(\mathbf{x} - \mathbf{x}_0\right)\right)\right]_{xk=-\infty}^{x_k=+\infty} dx_1 dx_2...$$

$$\times dx_{k-1} dx_{k+1}...dx_d \Delta t$$

$$= 0 . \tag{6.38}$$

In Eq. (6.38), when $l \neq k$, the delta function $\delta\left(x_k - (x_k)_0\right)$ is not affected by ∂_l and $\delta\left(x_k - (x_k)_0\right) = 0$ when evaluated at $x_k = \pm\infty$ since, again, the initial point $(x_2)_0$ is not at infinity. On the other hand, for $l = k$, the

Fluctuations without Memory 75

$(\partial/\partial x_k)\,\delta\,(x_k - (x_k)_0)$, or the first derivative of the delta function, is zero when x_k (which is at $\pm\infty$) is far from the initial point $(x_k)_0$. With Eqs. (6.37) and (6.38), the integral I_3 given by Eq. (6.32) is now written as,

$$I_3 = -\left(\frac{1}{2\pi}\right)^{d/2}\frac{1}{2}\int\limits_{-\infty}^{+\infty}[\partial_k\exp(i\mathbf{w}\cdot\mathbf{x})]\ \partial_l\,[b_{kl}\,(\mathbf{x},t)\,\delta\,(\mathbf{x}-\mathbf{x}_0)]\,d\mathbf{x}\Delta t$$

$$= -\left(\frac{1}{2\pi}\right)^{d/2}\frac{1}{2}\int\limits_{-\infty}^{+\infty}\exp(i\mathbf{w}\cdot\mathbf{x})\,(iw_k)\ \partial_l\,[b_{kl}\,(\mathbf{x},t)\,\delta\,(\mathbf{x}-\mathbf{x}_0)]\,d\mathbf{x}\Delta t$$

$$= -\left(\frac{1}{2\pi}\right)^{d/2}\frac{1}{2}\left\{\int\limits_{-\infty}^{+\infty}\partial_l\,[\exp(i\mathbf{w}\cdot\mathbf{x})\,(iw_k)\,b_{kl}\,(\mathbf{x},t)\,\delta\,(\mathbf{x}-\mathbf{x}_0)]\,d\mathbf{x}\Delta t\right.$$

$$\left.-\int\limits_{-\infty}^{+\infty}[\partial_l\exp(i\mathbf{w}\cdot\mathbf{x})]\,(iw_k)\,b_{kl}\,(\mathbf{x},t)\,\delta\,(\mathbf{x}-\mathbf{x}_0)\,d\mathbf{x}\Delta t\right\}. \tag{6.39}$$

The first integral in Eq. (6.39) can again be shown to be zero following the same arguments used in Eqs. (6.34) and (6.35). The I_3 would then be just the second integral which is,

$$I_3 = \left(\frac{1}{2\pi}\right)^{d/2}\frac{1}{2}\int\limits_{-\infty}^{+\infty}[\partial_l\exp(i\mathbf{w}\cdot\mathbf{x})]\,(iw_k)\,b_{kl}\,(\mathbf{x},t)\,\delta\,(\mathbf{x}-\mathbf{x}_0)\,d\mathbf{x}\Delta t$$

$$= -\left(\frac{1}{2\pi}\right)^{d/2}\frac{1}{2}\int\limits_{-\infty}^{+\infty}\exp(i\mathbf{w}\cdot\mathbf{x})\ w_k w_l\ b_{kl}\,(\mathbf{x},t)\,\delta\,(\mathbf{x}-\mathbf{x}_0)\,d\mathbf{x}\Delta t$$

$$= -\left(\frac{1}{2\pi}\right)^{d/2}\frac{1}{2}\exp(i\mathbf{w}\cdot\mathbf{x}_0)\,w_k w_l\,b_{kl}\,(\mathbf{x}_0,t)\,\Delta t\,, \tag{6.40}$$

where evaluation is facilitated by the Dirac delta function. Using Eqs. (6.30), (6.36), and (6.40) for I_1, I_2, and I_3 we now write the Fourier transform, Eq. (6.29), as

$$\mathcal{F}\,[P\,(\mathbf{x},t+\Delta t\mid\mathbf{x}_0,t)] = \left(\frac{1}{2\pi}\right)^{d/2}\exp(i\mathbf{w}\cdot\mathbf{x}_0)$$

$$\times\left\{1+\left[i\mathbf{w}\cdot\mathbf{A}\,(\mathbf{x}_0,t)-\frac{1}{2}w_k w_l\,b_{kl}\,(\mathbf{x}_0,t)\right]\Delta t\right\}, \tag{6.41}$$

which for $\Delta t \ll 1$ can be written as,

$$\mathcal{F}\left[P\left(\mathbf{x}, t + \Delta t \mid \mathbf{x}_0, t\right)\right] = \left(\frac{1}{2\pi}\right)^{d/2} \exp\left(i\mathbf{w} \cdot \mathbf{x}_0\right)$$

$$\times \exp\left\{\left[i\mathbf{w} \cdot \mathbf{A}\left(\mathbf{x}_0, t\right) - \frac{1}{2}w_k w_l\, b_{kl}\left(\mathbf{x}_0, t\right)\right]\Delta t\right\}$$

$$= \left(\frac{1}{2\pi}\right)^{d/2} \exp\{-\left(1/2\right)\mathbf{w}^T \mathbb{B}\left(\mathbf{x}_0, t\right)\mathbf{w}\Delta t$$

$$+ i\mathbf{w} \cdot \left[\mathbf{x}_0 + \mathbf{A}\left(\mathbf{x}_0, t\right)\Delta t\right]\}, \tag{6.42}$$

where $w_k b_{kl}\left(\mathbf{x}_0, t\right) w_l$, in matrix notation may be written as, $\mathbf{w}^T \mathbb{B}\left(\mathbf{x}_0, t\right)\mathbf{w}$.

We now evaluate the inverse Fourier transform of Eq. (6.42), i.e., we take, $\mathcal{F}^{-1}\mathcal{F}\left[P\left(\mathbf{x}, t + \Delta t \mid \mathbf{x}_0, t\right)\right] = P\left(\mathbf{x}, t + \Delta t \mid \mathbf{x}_0, t\right)$ which appears as,

$$P\left(\mathbf{x}, t + \Delta t \mid \mathbf{x}_0, t\right) = \left(\frac{1}{2\pi}\right)^d \int d\mathbf{w}\ \exp\left(-i\mathbf{w} \cdot \mathbf{x}\right)$$

$$\times \exp\left\{-\frac{1}{2}\mathbf{w}^T\, \mathbb{B}\left(\mathbf{x}_0, t\right)\Delta t\ \mathbf{w} + i\mathbf{w} \cdot \left[\mathbf{x}_0 + \mathbf{A}\left(\mathbf{x}_0, t\right)\Delta t\right]\right\}$$

$$= \left(\frac{1}{2\pi}\right)^d \int d\mathbf{w}\ \exp\left\{-\left(1/2\right)\mathbf{w}^T\, \mathbb{B}\left(\mathbf{x}_0, t\right)\Delta t\ \mathbf{w}\right.$$

$$\left. + i\mathbf{w} \cdot \left[\left(\mathbf{x}_0 - \mathbf{x}\right) + \mathbf{A}\left(\mathbf{x}_0, t\right)\Delta t\right]\right\}. \tag{6.43}$$

Equation (6.43) is a Gaussian integral over $d\mathbf{w}$ which can be evaluated to yield (see, e.g. Eq. (A-16) of [178]),

$$P\left(\mathbf{x}, t + \Delta t \mid \mathbf{x}_0, t\right) = \frac{1}{\left[\left(2\pi\right)^d \det\left(\mathbb{B}\left(\mathbf{x}_0, t\right)\Delta t\ \right)\right]^{1/2}}$$

$$\times \exp\left\{-\frac{1}{2}\left[\frac{\left(\mathbf{x}_0 - \mathbf{x}\right)}{\Delta t} + \mathbf{A}\left(\mathbf{x}_0, t\right)\right]^T\right.$$

$$\left. \times \left[\mathbb{B}\left(\mathbf{x}_0, t\right)\right]^{-1}\left[\frac{\left(\mathbf{x}_0 - \mathbf{x}\right)}{\Delta t} + \mathbf{A}\left(\mathbf{x}_0, t\right)\right]\Delta t\right\}, \tag{6.44}$$

which is a solution to the Fokker-Planck equation (6.25) for $\Delta t \ll 1$. For the one-dimensional case with drift coefficient $a\left(x, t\right)$ and constant diffusion coefficient D, Eq. (6.44) reduces to the form,

$$P\left(x,t+\Delta t\mid x_0,t\right)=\frac{1}{\left[2\pi D\Delta t\;\right]^{1/2}}\exp\left\{-\frac{1}{2D}\left[\frac{\left(x-x_0\right)}{\Delta t}-a\left(x_0,t\right)\right]^2\Delta t\right\},$$

(6.45)

which is a short-time solution to the one-dimensional Fokker-Planck equation. Note that the diffusion coefficient in Eq. (6.45) differs by a factor of 2, compared for example with Eq. (4.55) of [181], in view of the starting Fokker-Planck equation, Eq. (6.25) which has a factor $1/2$ in the second term.

6.3 Path Integral for the Fokker-Planck Equation

The solution $P\left(\mathbf{x},t\mid\mathbf{x}_0,t_0\right)$ of the Fokker-Planck equation (6.25) can also be expressed as a path integral for any finite time interval [181; 210], $\tau=t-t_0$. This is done using Eq. (6.45) and the Chapman-Kolmogorov equation for Markov processes given by,

$$P\left(x_3,t_3\mid x_1,t_1\right)=\int P\left(x_3,t_3\mid x_2,t_2\right)P\left(x_2,t_2\mid x_1,t_1\right)dx_2.$$

(6.46)

We proceed by dividing the time τ into N sub-intervals, i.e., $\tau/N=\Delta t_j=t_j-t_{j-1}$, where $j=1,2,...,N$, such that the final time is $t=t_N$. Consider now the one-dimensional case where we define the position at time t_j as, $x_j=x\left(t_j\right)$ with the endpoint, $x=x_N$. Given these small time intervals, we apply Eq. (6.46) repeatedly to express $P\left(x,t\mid x_0,t_0\right)$ as,

$$P\left(x,t\mid x_0,t_0\right)=\int\cdots\int P\left(x_N,t_N\mid x_{N-1},t_{N-1}\right)$$

$$\times P\left(x_{N-1},t_{N-1}\mid x_{N-2},t_{N-2}\right)$$

$$\times...\;P\left(x_{j+1},t_{j+1}\mid x_j,t_j\right)P\left(x_j,t_j\mid x_{j-1},t_{j-1}\right)\;...$$

$$\times P\left(x_2,t_2\mid x_1,t_1\right)P\left(x_1,t_1\mid x_0,t_0\right)\;dx_1dx_2...dx_{N-1}.$$

(6.47)

Since each probability density $P\left(x_j,t_j\mid x_{j-1},t_{j-1}\right)$ involves small intervals of time, $\Delta t_j=t_j-t_{j-1}$, we can substitute the expression given by Eq. (6.45) to write Eq. (6.47) as,

$$
P\left(x, t \mid x_0, t_0\right) = \int \prod_{j=1}^{N} \frac{1}{[2\pi D \Delta t_j]^{1/2}}
$$

$$
\times \exp\left\{-\frac{1}{2D}\left[\frac{\Delta x_j}{\Delta t_j} - a\left(x_{j-1}, t_j\right)\right]^2 \Delta t_j\right\} \prod_{j=1}^{N-1} dx_j
$$

$$
= \int \exp\left[-\sum_{j=1}^{N} L_j \Delta t_j\right] \prod_{j=1}^{N} \frac{1}{[2\pi D \Delta t_j]^{1/2}} \prod_{j=1}^{N-1} dx_j,
$$

$$(6.48)$$

where $\Delta x_j = x_j - x_{j-1}$, and

$$
L_j = \frac{1}{2D}\left[\frac{\Delta x_j}{\Delta t_j} - a\left(x_{j-1}, t_j\right)\right]^2. \tag{6.49}
$$

In the limit, $N \to \infty$ and $\Delta t_j \to 0$, we have $\sum L_j \Delta t_j \to S = \int L dt$, which is our effective action, and Eq. (6.48) can be written symbolically as the path integral,

$$
P\left(x, t \mid x_0, t_0\right) = \int \exp\left\{-\frac{1}{2D} \int \left[\frac{dx}{dt} - a\left(x, t\right)\right]^2 dt\right\} \mathcal{D}[x]
$$

$$
= \int \exp\left(-S\right) \mathcal{D}[x]. \tag{6.50}
$$

6.4   One-Dimensional Random Walk

We shall now illustrate the use of white noise calculus [106; 133; 163] in evaluating the probability function by considering the one-dimensional random walk problem. We begin with the Wiener representation of the random walk along the x-axis which starts at x_0 and ends at x_1 given by,

$$
P(x_1, x_0; L) = \int \exp\left[-\frac{1}{2l} \int_0^L \left(\frac{dx}{ds}\right)^2 ds\right] \mathcal{D}[x]. \tag{6.51}
$$

In this probability function, each step is denoted by l and the total number of steps N is such that $Nl = L$. The probability function Eq. (6.51) can be cast in the language of white noise [196] by parametrizing the paths as,

$$x(L) = x_0 + \sqrt{2l}\, B(L)$$

$$= x_0 + \sqrt{2l} \int_0^L \omega(s)\, ds \,, \tag{6.52}$$

where $B(s)$ is a Brownian motion parametrized by s, $0 \le s \le L$, and $\omega = dB/ds$, is the corresponding white noise variable. With Eq. (6.52), the exponential in Eq. (6.51) becomes,

$$\exp\left[-\frac{1}{2l} \int_0^L \left(\frac{dx}{ds} \right)^2 ds \right] = \exp\left[-\int_0^L \omega(s)^2\, ds \right], \tag{6.53}$$

where $(dx/ds) = \sqrt{2l}\, \omega$. Since the integrand in Eq. (6.51) is now expressed as a white noise functional, the integral over the paths $\mathcal{D}[x]$ becomes an integral over the Gaussian white noise measure $d\mu(\omega)$ which has the form,

$$d\mu(\omega) = N_\omega \exp\left[-\frac{1}{2} \int_0^L \omega(s)^2\, ds \right] d^\infty \omega \tag{6.54}$$

from Eq. (2.7). More appropriately, the $\mathcal{D}[x]$ (where $\mathcal{D}[x] = d^\infty x$, in the context of path integrals [86]) is replaced by $N_\omega\, d^\infty \omega = \exp\left[(1/2) \int \omega(s)^2\, ds \right] d\mu(\omega)$. With the exponential multiplying $d\mu(\omega)$, we are led to a modification of Eq. (6.53) and shall, therefore, consider the white noise functional,

$$I_0 = N \exp\left(-\int_0^L \omega(s)^2\, ds \right) \exp\left(\frac{1}{2} \int_0^L \omega(s)^2\, ds \right)$$

$$= N \exp\left(-\frac{1}{2} \int_0^L \omega(s)^2\, ds \right), \tag{6.55}$$

where N is an appropriate normalization factor. To incorporate the end-point x_1, we use the Donsker delta function [106; 140] $\delta(x(L) - x_1)$, to pin down the paths $x(s)$, where $x(L)$ is given by Eq. (6.52). This, together with the white noise functional Eq. (6.55), enables us to write the probability function, Eq. (6.51), as

$$P(x_1, x_0; L) = N \int \exp\left(-\frac{1}{2}\int\limits_0^L \omega(s)^2\, ds\right)$$

$$\times \delta\left(x_0 + \sqrt{2l}\int\limits_0^L \omega(s)\, ds - x_1\right) d\mu(\omega). \qquad (6.56)$$

To evaluate Eq. (6.56), we may use the Fourier representation of the δ-function, i.e.,

$$P(x_1, x_0; L) = \frac{1}{2\pi}\int\limits_{-\infty}^{+\infty} d\lambda \exp\left[i\lambda(x_0 - x_1)\right]$$

$$\times N \int \exp\left(i\lambda\sqrt{2l}\int\limits_0^L \omega(s)\, ds\right)$$

$$\times \exp\left(-\frac{1}{2}\int\limits_0^L \omega(s)^2\, ds\right) d\mu(\omega). \qquad (6.57)$$

The integration over $d\mu(\omega)$ can be done by noticing that, from white noise calculus [106; 133; 163], the T-transform of I_0, Eq. (6.55), is given by,

$$TI_0(\xi) = \int \exp\left(i\int \omega\xi\, ds\right) I_0(\omega)\, d\mu(\omega)$$

$$= \exp\left(-\frac{1}{4}\int\limits_0^L \xi^2\, ds\right), \qquad (6.58)$$

where $N^{-1} = \int \exp\left[-(1/2)\int \omega(s)^2\, ds\right] d\mu(\omega)$ is the normalization. If we let $\xi = \lambda\sqrt{2l}$ in Eq. (6.57), then the integration over $d\mu(\omega)$ is just $TI_0(\xi)$, Eq. (6.58). The probability function, Eq. (6.57), therefore becomes,

$$P(x_1, x_0; L) = \frac{1}{2\pi}\int\limits_{-\infty}^{+\infty} \exp\left[-(lL/2)\lambda^2 + i(x_0 - x_1)\lambda\right] d\lambda. \qquad (6.59)$$

What remains is a Gaussian integral over λ, and we obtain the familiar result for a one-dimensional random walk starting from x_0 and ending at x_1 [55],

$$P(x_1, x_0; L) = \sqrt{1/2\pi N l^2} \exp\left[-\left(1/2N l^2\right)(x_0 - x_1)^2\right], \qquad (6.60)$$

where $Nl = L$.

Chapter 7

Neurophysics

Recent experimental results in neuroscience indicate that major factors affecting the interactions between neurons include: the location, geometry, synaptic molecular dynamics and chemical composition of the neuron, and transneuronal ion currents. The analytical description of neuronal behavior is extremely challenging because there are about 10^{11} neurons in the entire human brain [2; 94; 115; 202; 216]. These are of different shapes and types, and are generally arranged in layers and clusters or constellations, whose geometry could be related to the functions of each neuronal cluster. For example, while there are clusters arranged in lattice configuration, there are others appearing as almost flat two-dimensional plates. Still other clusters seem randomly arranged throughout a cortical layer. Furthermore, each neuron has a system of dendrites and axons with synapses linking it to around 10^4 other neurons. Intercellular communication through synapses is then initiated by the transmission of electrical signals generated when a neuron 'fires', or releases ionic currents to revert to equilibrium relative membrane potentials. Experimental results show a spatial distribution of neuronal activity during performance of particular tasks: grammar processing shows activity in frontal regions of the left hemisphere of the brain, while semantic processing shows activity in the posterior lateral regions of both left and right hemispheres. Also of special interest for pedagogical applications [164] is the Triple Code Model. Results show that visual perception of a number, say 5, involves the fusiform gyrus, hearing the number spoken activates the perisylvian area, and understanding that five is greater than one involves the interparietal lobes. Finally, neurons are dynamic biological powerhouses with the ability to grow new synapses, or retract dysfunctional ones, and to synthesize proteins, in the soma or even at synaptic sites, on demand [15], depending on identified needs and tasks.

Finer spatio-temporal mapping of the brain is being made possible by continuing advances in medical imaging such as enhanced postiron emission tomography (PET) scans and functional magnetic resonance imaging (fMRI). Another development, optical topography, allows subjects to move, unlike the PET where the subject should be immobile for 40 minutes or more for sharper image resolution. The technique developed by Koizumi *et al.* [130] targets higher-order brain function analysis by trans-cranial dynamic near-infrared spectroscopy imaging. Electro- and magneto-encephalography with superconducting quantum interference devices (SQUIDS) allow high resolution temporal mapping of brain activity (of the order of 10^{-3}s). Even with these advances, however, much remains to be understood because, aside from the complexity of brain structure, huge numbers of variables are involved in carrying out functions and tasks. Extensive work has been done including theoretical models to accommodate various features of the brain. A popular approach is the use of neural networks. Each neuron is treated as a simple computer and the brain as a large composite computer [108]. There are limitations to this approach, notably, unlike two-state computer bits that are either "on" or "off", a single neuron can be activated through a continuum of states, and links with a large number of other neurons responding with their own spectrum of neuronal states. Another model views the brain as a nonlinear nonequilibrium open system with infinite degrees of freedom. Macroscopic patterns in aggregates of neurons, instead of the activity of single neurons, are considered. The evolution and dynamics of the patterns are then interpreted as cognitive flows. Such a picture treats the brain as a pattern-forming, self-organized, dynamical complex system [120; 209].

Considering the complexity and breadth in scope of brain studies, it remains important to investigate different models, approaches and modern mathematical tools for possible insights that could lead to a deeper understanding of how the brain works [138]. Indeed, better models that are explanatory, quantitative, and predictive could have tremendous implications in the health sciences, education and learning, and the dynamics of social systems and communities. This motivates us to explore a white noise path integral approach as a mathematical framework that may be rich enough in structure to accommodate different features of the brain: geometry and global topology, structure, large N-dimensional configuration spaces, and dynamical fluctuations typical of biological systems.

We start with the parametrization idea of neuron firing introduced by K. Yasue *et al.* [214], identifying the dynamical random variables as the

Table 7.1. A qualitative mapping of mathematical models according to scope and descriptive details from a single neuron to populations of neurons.

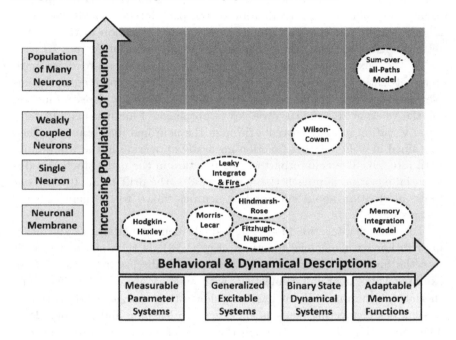

relative membrane potentials and currents. However, we employ the path integral method to investigate the stochastic dynamics of the system. The path integral, or sum-over-all-histories approach, is a most natural framework for treating infinite-dimensional systems. This has been recognized earlier, for example, by M. Turvey [187], who suggested the use of the path integral for modelling how the cerebellum selects and shapes a control signal from background neuronal excitation. The path integral in the spin coherent state representation has been used in a master equation for neural networks by Ohira and Cowan [166]. Moreover, the path integral also has the added advantage of being able to deal with nontrivial topological features [21] that may arise in modelling brain functions.

In this chapter, we adopt a different use of the Feynman path integral to model features of brain processes (see, e.g., rightmost column of Table 7.1. This table is adapted, with modification, from [53]). First of all, the path integral is interpreted in the sense of Hida and Streit [196], i.e., as an average over a Gaussian white noise measure of well-defined distributions on the infinite dimensional coordinate system formed by the random white noise variables, $\omega(t) = dB/dt, \ t \in \mathbf{R}$. Here, B denotes the usual Brownian motion. Evaluation of the path integral yields the probability density function describing the stochastic variation in time of the relative membrane potential of the neuron as discussed in Section 7.1.

In the next section, we shall briefly describe the white noise path parametrization approach of Hida and Streit. This is followed by a simple modelling of neuronal activity for which a probability density function for the relevant dynamical variables is obtained. Fluctuations and temporal variation in ionic currents through the neuronal membrane are then contained in a modulating function naturally appearing in the white noise path integral. The counterpart of this function in the usual stochastic differential equation approach (Langevin type) is the drift term. It also contains information about surrounding neurons linked to the neuron being considered.

For this chapter, we first model single neuron activities. The parametrization with memory functions, Eq. (1.13), is then shown to facilitate the investigation of neuronal systems. We discuss in Section 7.3 an example where neuronal connections with memory are modelled [24]. Furthermore, experimentally, it has been shown that changes in the strength of synapses connecting neurons are crucial to learning and memory [183]. The conduction of signals through the maze of axons, synapses and dendrites from an initiator neuron to a target receptor neuron, therefore, can be modelled by an ensemble of fractional Brownian paths using Eq. (3.14) as a memory function. Some remarks are also given towards the end of the chapter on the issue of how the brain encodes and decodes information with a model that utilizes the memory functions, Eqs. (3.17) and (3.20).

7.1 Modelling Single Neuron Activities

7.1.1 *Relative Membrane Potential as Random Variable*

The interior of a neuron is normally separated from the outside by a 5-nm-thick neuronal membrane that serves as a barrier preventing the inflow of certain substances into the extracellular fluid. There is, however, a diffusive movement of charged ions across the neuronal membrane, and an electrical field arising from voltage differences across the membrane. In the so-called resting neuron, the voltage difference arises from the cytosol along the inner surface of the membrane negatively charged compared with the outside. We can then designate the voltage across the neuronal membrane at any given time t as $V_m(t)$. This is referred to as the membrane potential. Whenever the membrane potential crosses a threshold voltage $V_T(t)$, the neuron fires and sends an impulse that propagates along its axon. To characterize the firing of neurons we can then consider a variable $v(t)$ which is the difference of two voltages:

$$v(t) = V_m(t) - V_T(t). \qquad (7.1)$$

This is called the relative membrane potential [214]. A neuron fires and is active when, $v(t) > 0$, and is quiescent when $v(t) < 0$.

Let us look at a single neuron and consider its fluctuations so that the random variable $v(t)$ obeys the stochastic differential equation (2.1), and the corresponding Langevin equation for $v(t)$ (see, e.g., Eq. (2.2)) is given by,

$$\frac{dv(t)}{dt} = a(v,t) + \sqrt{D}\,\omega(t), \qquad (7.2)$$

where $a(v,t)$ and D are drift and diffusion coefficients, respectively, and $\omega(t)$ is a Gaussian white noise term.

In the simple case of a single neuron characterized by $v(t)$, Eq. (7.1), the stochastic variation of the random variable $v(t)$ between an initial value $v(0) = v_0$, and final value $v(t) = v$ can be characterized by the conditional probability density $P(v,t \mid v_0,t_0)$. This density function satisfies the Fokker-Planck equation corresponding to the stochastic differential equation, Eq. (7.2). We have

$$\frac{\partial}{\partial t}P(v,t \mid v_0,t_0) = -\frac{\partial}{\partial v}[a(v,t)P(v,t \mid v_0,t_0)]$$

$$+\frac{1}{2}D\frac{\partial^2}{\partial v^2}P(v,t \mid v_0,t_0). \qquad (7.3)$$

We shall now proceed to solve Eq. (7.3) and obtain exact expressions for the conditional probability density $P(v,t \mid v_0,t_0)$.

7.1.2 *Path Integral for a Single Neuron*

One can solve the differential equation (7.3) directly. However, alternatively, as a springboard for nontrivial applications, we obtain the conditional probability density $P(v_1,t \mid v_0,t_0)$ by expressing it as a path integral (see Eq. (6.50)) [182],

$$P\left(v_1, t \mid v_0, t_0\right) = \int \exp\{-S[v(t)]\}\mathcal{D}\left[v(t)\right] . \qquad (7.4)$$

The integration over $\mathcal{D}\left[v(t)\right]$ is over all possible histories of the fluctuating relative membrane potential $v(t)$ from the initial value v_0 to the specified terminal value v_1. The effective action $S[v(t)]$ is given by (setting $t_0 = 0$),

$$S = \frac{1}{2D} \int_0^t [\dot{v} - a(v, \tau)]^2 d\tau , \qquad (7.5)$$

where $\dot{v} = dv/d\tau$. To evaluate Eq. (7.4), we make use of the general method first introduced by Hida and Streit for the Feynman path integral [196], also naturally applicable to the Feynman-Kac integral. First, starting from v_0, we parametrize the histories in terms of the Brownian motion $B\left(t\right)$,

$$v(t) = v_0 + \sqrt{2D}B\left(t\right) , \qquad (7.6)$$

and, with v_0 fixed, the time derivative of $v(t)$ is given by,

$$\dot{v} = \sqrt{2D}\frac{dB\left(\tau\right)}{d\tau}$$
$$= \sqrt{2D}\omega(\tau) , \qquad (7.7)$$

where the Gaussian white noise variable $\omega(\tau) = dB/d\tau$. With Eq. (7.7), the action (7.5) can now be written as,

$$S = \int_0^t \omega^2(\tau) \, d\tau - \sqrt{\frac{2}{D}} \int_0^t \omega(\tau) \, a(v, \tau) \, d\tau$$

$$+ \frac{1}{2D} \int_0^t a^2(v, \tau) d\tau . \qquad (7.8)$$

If we take the simple case of a time-dependent drift coefficient, i.e., $a(v, t) = \xi\left(t\right)$, we obtain the following exponential in (7.4) as a white noise functional,

$$\exp\{-S[v(t)]\} = \exp\left(-\int_0^t \omega^2(\tau)\,d\tau + \sqrt{\frac{2}{D}}\int_0^t \omega(\tau)\xi(t)\,d\tau\right)$$

$$\times \exp\left(-\frac{1}{2D}\int_0^t \xi(t)^2\,d\tau\right). \tag{7.9}$$

Note that the parametrization (7.6) does not fix the final endpoint at $v(t) = v_1$. We can, however, fix the endpoint by using the Donsker delta function,

$$\delta\left(v(t) - v_1\right) = \delta\left(v_0 + \sqrt{2D}B(t) - v_1\right)$$

$$= \delta\left(v_0 - v_1 + \sqrt{2D}\int_0^t \omega(\tau)\,d\tau\right), \tag{7.10}$$

where Eq. (7.6) was used for $v(t)$ and $B(t) = \int_0^t \omega(\tau)\,d\tau$. Furthermore, since the integrand in Eq. (7.4) is now a white noise functional given by Eq. (7.9), the integration over the paths $\mathcal{D}[v(t)]$ or $d^\infty v$ is transformed into an integration over the Gaussian white noise measure, $d\mu(\omega)$. In particular, from Eq. (2.7) we have, $d^\infty v \to N_\omega d^\infty \omega = \exp\left[\frac{1}{2}\int \omega(s)^2\,ds\right]d\mu(\omega)$, with N_ω a normalization factor. Using this measure, Eqs. (7.9) and (7.10), the path integral for $P(v_1, t \mid v_0, t_0)$, Eq. (7.4), can now be written as,

$$P(v_1, t \mid v_0, t_0) = \int I_0 \exp\left(\sqrt{\frac{2}{D}}\int_0^t \omega(\tau)\xi(t)\,d\tau - \frac{1}{2D}\int_0^t \xi(t)^2\,d\tau\right)$$

$$\times \delta\left(v_0 - v_1 + \sqrt{2D}\int_0^t \omega(\tau)\,d\tau\right)d\mu(\omega). \tag{7.11}$$

In Eq. (7.11), I_0 is a Gauss kernel,

$$I_0 = N \exp\left(-\frac{1}{2}\int_0^t \omega^2(\tau)d\tau\right), \tag{7.12}$$

obtained from the exponential in $N_\omega d^\infty \omega$ and the first term in Eq. (7.9), with N an appropriate normalization. Equation (7.11) can further be simplified by expressing the Donsker delta function in terms of its Fourier representation to get,

$$P\left(v_1, t \mid v_0, t_0\right) = \exp\left(-\frac{1}{2D}\int_0^t \xi^2(t)d\tau\right)$$

$$\times \frac{1}{2\pi}\int_{-\infty}^{+\infty} dk \exp\left[ik\left(v_0 - v_1\right)\right]$$

$$\times \int I_0 \exp\left\{i\int_0^t \omega(\tau)\left[k\sqrt{2D} - i\sqrt{\frac{2}{D}}\xi(t)\right]d\tau\right\} d\mu\left(\omega\right).$$

$$(7.13)$$

Denoting, $\xi' = k\sqrt{2D} - i\sqrt{2/D}\xi(t)$, the integration over $d\mu\left(\omega\right)$ is just the T-transform of the Gauss kernel I_0 (see, e.g., Eqs. (2.8) and (2.27)), i.e.,

$$TI_0\left(\xi'\right) = \int I_0 \exp\left[i\int_0^t \omega(\tau)\xi'(\tau)d\tau\right] d\mu\left(\omega\right)$$

$$= \exp\left(-\frac{1}{4}\int_0^t \xi'^2(\tau)d\tau\right)$$

$$= \exp\left(-\frac{k^2 Dt}{2} + ik\int_0^t \xi(\tau)d\tau + \frac{1}{2D}\int_0^t \xi^2(\tau)d\tau\right). \quad (7.14)$$

Using Eq. (7.14) in Eq. (7.13), we obtain the expression for the conditional probability,

$$P\left(v_1, t \mid v_0, t_0\right) = \frac{1}{2\pi}\int_{-\infty}^{+\infty} dk \exp\left(-k^2 Dt/2\right)$$

$$\times \exp\left\{ik\left[\left(v_0 - v_1\right) + \int_0^t \xi(\tau)d\tau\right]\right\}. \quad (7.15)$$

The remaining integral over dk is a Gaussian integral which can be evaluated to yield,

$$P\left(v_1, t \mid v_0, t_0\right) = \sqrt{\frac{1}{2\pi Dt}} \exp\left[-\frac{1}{2Dt}\left(v_0 - v_1\right)^2\right]$$

$$\times \exp\left[-\frac{\left(v_0 - v_1\right)}{Dt} \int\limits_0^t \xi(\tau)d\tau\right]$$

$$\times \exp\left\{-\frac{1}{2Dt}\left[\int\limits_0^t \xi(\tau)d\tau\right]^2\right\}. \tag{7.16}$$

We have thus obtained an expression for the conditional probability density function for the relative membrane potential involved in neural activity for the case when there is a time-dependent drift. With $v\left(t\right)$ being a voltage difference that triggers the firing of the neuron, the drift coefficient can be interpreted as modulating the variation of total current crossing the neuronal membrane. In an explicit solution for $P\left(v_1, t \mid v_0, t_0\right)$ the time-dependent drift coefficient $a\left(t\right) = \xi(t)$ can be chosen to model certain features of brain processes. Periodic oscillations of cellular currents can be described by taking $a(t) = \sigma\cos\left(\omega t\right)$, a rapidly increasing change in current by $\xi(t) = \lambda\exp\left(ct\right)$, or by the polynomial form, $\xi(t) = \alpha t^s$, where σ, λ, and α are biochemical parameters. Such a polynomial dependence in time of the drift can already strongly damp the probability density function for the firing of a neuron.

7.1.3 *Voltage and Time-Dependent Current Modulation*

We now consider a drift, or a membrane current modulating function, which is both voltage and time-dependent. In particular, we take $a(v, \tau)$ to be,

$$a(v, \tau) = v\,\xi\left(\tau\right), \tag{7.17}$$

which implies that modulation of the neuronal membrane current now depends on the relative membrane potential v defined by Eq. (7.1). For this case, we would have interaction terms in the action of the path integral,

Eq. (7.5), of the form,

$$S = \frac{1}{2D} \int_{t_0}^{t} \left[\dot{v}^2 - 2\dot{v}\, a(v,\tau) + a^2(v,\tau) \right] d\tau$$

$$= \frac{1}{2D} \int_{t_0}^{t} \left[\dot{v}^2 - 2\dot{v}v\xi + v^2\xi^2 \right] d\tau \,, \tag{7.18}$$

with t_0 and t the initial and final time, respectively. The middle term in Eq. (7.18) can be written as,

$$\int_{t_0}^{t} 2\dot{v}v\xi d\tau = \int_{t_0}^{t} 2\frac{dv}{d\tau}v\xi d\tau$$

$$= \int_{t_0}^{t} \frac{d\left(v^2\right)}{d\tau}\xi$$

$$= \int_{t_0}^{t} \frac{d}{d\tau}\left(v^2\xi\right) d\tau - \int_{t_0}^{t} v^2 \frac{d\xi}{d\tau} d\tau$$

$$= \left[v^2(t)\,\xi(t) - v^2(t_0)\,\xi(t_0) \right] - \int_{t_0}^{t} v^2\dot{\xi} d\tau \,. \tag{7.19}$$

Putting back Eq. (7.19) into Eq. (7.18), we get an action of the form,

$$S = S_0 + S_1 \,, \tag{7.20}$$

where S_0 is a constant given by,

$$S_0 = \frac{1}{2D} \left[v_1^2 \xi_1 - v_0^2 \xi_0 \right] \,, \tag{7.21}$$

with values of the variables at the initial time, $v_0 = v(t_0)$ and $\xi_0 = \xi(t_0)$, and at the final time, $v_1 = v(t)$ and $\xi_1 = \xi(t)$. The S_1, on the other hand, is given by,

$$S_1 = \frac{1}{2D} \int_{t_0}^{t} \left[\dot{v}^2 + \Omega^2(\tau)\, v^2 \right] d\tau \,, \tag{7.22}$$

where $\Omega^2(\tau)$ is,

$$\Omega^2(\tau) = \dot{\xi} + \xi^2 . \qquad (7.23)$$

We now observe that S_1 is similar to an action for a time-dependent harmonic oscillator. The conditional probability density, Eq. (7.4) is now given by,

$$P(v_1, t \mid v_0, t_0) = \exp(-S_0) \, P_{osc}(v_1, t \mid v_0, t_0) , \qquad (7.24)$$

where $P_{osc}(v, t \mid v_0, t_0)$ is the path integral for the time-dependent harmonic oscillator of the form,

$$P_{osc}(v_1, t \mid v_0, t_0) = \int \exp\left\{ -\frac{1}{2D} \int_{t_0}^{t} \left[\dot{v}^2 + \Omega^2(\tau) \, v^2 \right] d\tau \right\} \mathcal{D}[v(t)] .$$
$$(7.25)$$

Path integration for the time-dependent harmonic oscillator in quantum mechanics has already been done (see, e.g., [121]). In the context of white noise path integrals, the time-dependent harmonic oscillator can also be found in [100]. Applying transformations for time and mass, where $t \to -it$, and $m \to 1/D$ in the quantum mechanical propagator [121] and setting $\hbar = 1$, we obtain the solution for the probability density, Eq. (7.25), given by,

$$P_{osc}(v_1, t \mid v_0, t_0) = \left(\frac{\sqrt{\dot{\gamma}\dot{\gamma}_0}}{2\pi D \sin \phi(t, t_0)} \right)^{1/2} \exp\left[\frac{1}{2D} \left(\frac{\dot{s}v_1^2}{s} - \frac{\dot{s}_0 v_0^2}{s_0} \right) \right]$$

$$\times \exp\left\{ \frac{1}{2D \sin \phi(t, t_0)} \left[\left(\dot{\gamma}v_1^2 + \dot{\gamma}_0 v_0^2 \right) \cos \phi(t, t_0) \right.\right.$$

$$\left.\left. -2\sqrt{\dot{\gamma}\dot{\gamma}_0} v_0 v_1 \right] \right\} , \qquad (7.26)$$

where dots over s and γ denote differentiation with respect to time and

$$\phi(t, t_0) = \gamma(t) - \gamma(t_0) . \qquad (7.27)$$

The $s(t)$ and $\gamma(t)$ represent the amplitude and phase, respectively, in a classical time-dependent oscillator [121]. This means that if $\eta(t)$ denotes displacement, the equation of motion for the harmonic oscillator is,

$$\ddot{\eta}(t) + \Omega^2(t)\,\eta(t) = 0\,, \qquad (7.28)$$

which has a solution for real $\Omega(t)$ given by,

$$\eta(t) = s(t)\exp[i\gamma(t)]\,. \qquad (7.29)$$

For the case when Ω is a real positive constant, the amplitude and phase become, $s = \sqrt{m/\Omega}$ and $\gamma(t) = \Omega t$, respectively, where m is mass [121].

In Eq. (7.26), $s(t)$ and $\gamma(t)$ also obey the differential equations,

$$\ddot{s} - C^2 s^{-3} + \Omega^2 s = 0\,, \qquad (7.30)$$

$$\dot{\gamma}s^2 = C\,, \qquad (7.31)$$

where C is an arbitrary constant. We will next discuss a solution to Eqs. (7.30) and (7.31) giving specific forms for $s(t)$ and $\gamma(t)$ which can be used in Eq. (7.26).

7.1.4   *Membrane Current Modulation:* $a(v,\tau) = b\,(v/\tau)$

Explicit exact expressions for the conditional probability density can be obtained from Eq. (7.26). As an example, let us consider a drift coefficient, or current modulation coefficient of the form,

$$a(v,\tau) = b\,\frac{v}{\tau}\,, \qquad (7.32)$$

where we have taken the time dependence in Eq. (7.17) to be,

$$\xi(\tau) = \frac{b}{\tau}\,. \qquad (7.33)$$

Equation (7.32) implies that modulation of the flow of ionic current across the neuronal membrane depends on the relative membrane potential, Eq. (7.1), but this dependence on v diminishes as elapsed time τ increases. For this case, explicit expressions for $s(t)$ and $\gamma(t)$ can be obtained and used in Eq. (7.26). With Eq. (7.33), the time-dependent frequency Eq. (7.23) becomes,

$$\Omega^2(\tau) = \frac{a^2}{\tau^2}\,, \qquad (7.34)$$

with $a^2 = b(b-1)$. A solution to Eq. (7.30) would be [79],

$$s^2(\tau) = \alpha\tau,$$ (7.35)

where

$$\alpha = \frac{C}{\left(a^2 - \frac{1}{4}\right)^{1/2}}.$$ (7.36)

From these, we get expressions for s, $\dot{s}$, γ, $\dot{\gamma}$ and $\phi(t, t_0)$ needed in Eq. (7.26). From Eq. (7.35), we get,

$$\dot{s} = \frac{d}{d\tau}(\alpha\tau)^{1/2} = \sqrt{\frac{\alpha}{4\tau}}.$$ (7.37)

On the other hand, Eqs. (7.31) and (7.35) provide as the form (taking $C = 1$),

$$\dot{\gamma} = \frac{1}{\alpha\tau}.$$ (7.38)

From this, we obtain,

$$\gamma = \frac{1}{\alpha}\int \frac{d\tau}{\tau} = \frac{1}{\alpha}\ln t.$$ (7.39)

Finally, from Eqs. (7.27) and (7.39) we get,

$$\phi(t, t_0) = \frac{1}{\alpha}(\ln t - \ln t_0) = \ln\left(\frac{t}{t_0}\right)^{1/\alpha}.$$ (7.40)

Hence, using Eqs. (7.35), (7.37)–(7.40), then Eq. (7.26) acquires the form,

$$P_{osc}(v_1, t \mid v_0, t_0) = \left(2\pi\alpha D\sqrt{tt_0}\sin\phi(t, t_0)\right)^{-1/2}\exp\left[\frac{1}{4D}\left(\frac{v_1^2}{t} - \frac{v_0^2}{t_0}\right)\right]$$

$$\times \exp\left\{\frac{1}{2\alpha D\sin\phi(t,t_0)}\left[\left(\frac{v_1^2}{t} + \frac{v_0^2}{t_0}\right)\cos\phi(t,t_0)\right.\right.$$

$$\left.\left.-\frac{2v_0v_1}{\sqrt{tt_0}}\right]\right\}.$$ (7.41)

Equation (7.41) together with Eq. (7.21) yield a conditional probability density, Eq. (7.24), of the form,

$$P\left(v_1, t \mid v_0, t_0\right) = \left(2\pi\alpha D\sqrt{tt_0}\sin\phi\left(t, t_0\right)\right)^{-1/2}\exp\left(-\frac{1}{2D}\left[v_1^2\xi_1 - v_0^2\xi_0\right]\right)$$

$$\times\exp\left[\frac{1}{4D}\left(\frac{v_1^2}{t} - \frac{v_0^2}{t_0}\right)\right]$$

$$\times\exp\left\{\frac{1}{2\alpha D\sin\phi\left(t, t_0\right)}\left[\left(\frac{v_1^2}{t} + \frac{v_0^2}{t_0}\right)\cos\phi\left(t, t_0\right)\right.\right.$$

$$\left.\left. -\frac{2v_0 v_1}{\sqrt{tt_0}}\right]\right\}. \tag{7.42}$$

Once a probability density function is obtained, a theoretical model is deemed essentially solved since other properties may be derived from it. In view of the various analytical approaches in understanding neuronal activities, however, a convergence of theory and experiment needs to be worked on [56; 120; 131]. Continuing technological developments, however, should be able to guide theoretical, computational, and mathematical work in neuroscience concerning validity of results, such as Eq. (7.42), for correspondence with the behavior of real neurons.

7.1.5 *N Neurons*

From individual neurons, how would we extend to groups of neurons? We can designate each specific physical location of a neuron in the brain as a point $\mathbf{x} \in \mathbf{R}^3$ in the usual three-dimensional space. For example, in spherical coordinates, if we take the origin to be the geometric center of the brain, then $\mathbf{x} = \{r, \theta, \varphi\}, 0 \leq r < r_b$, (with r_b indicating the radial extent of the brain), $-\pi \leq \theta < \pi, 0 \leq \varphi < 2\pi$. The location of the neuron also specifies the origin of the impulse it generates. To locate the three-dimensional physical position of N $(N \approx 10^{10})$ neurons, we take $\mathbf{x}_i$, where in spherical coordinates, $\mathbf{x}_i = \{r, \theta, \varphi\}_i$, $i = 1, 2, ..., N$. Clusters of neurons may form a domain, Ω. Neurons at $\mathbf{x}_i \in \Omega_a$ may be responsible for hearing, while those at $\mathbf{x}_j \in \Omega_b$, with $\Omega_a, \Omega_b \in \mathbf{R}^3$ are for sight, and so on. With this, for each neuron at $\mathbf{x}_i$ there is also associated a relative membrane potential $v_i(t)$ defined as in Eq. (7.1). Furthermore, the nature of the impulse produced by any neuron depends on the stochastic temporal variation of the potential difference between intracellular and external ionic concentrations. Thus, the activity of the ith neuron and its interaction with other neurons can be described as $\boldsymbol{\eta}_i = \boldsymbol{\eta}_i(\mathbf{x}_i, v_i(t), t)$. This also leads to generalization to N neurons with path integrals in large-N dimensions. Symbolically, such path integral may be written as,

$$P\left(\widehat{v},t \mid \widehat{v}_0,t_0\right) = \int e^{-S(\widehat{v})} \mathcal{D}\left[\widehat{v}\right], \qquad (7.43)$$

where $\widehat{v}= (v_1, v_2, ..., v_N)$ and an effective action, $S\left(\widehat{v}\right) = \int \mathcal{L}\, dt$, with a Lagrangian,

$$\mathcal{L} = \frac{1}{2} \left(\frac{d\widehat{v}}{dt} - \mathbf{A}\left(\widehat{v},t\right)\right)^T \mathbb{B}\left(\widehat{v},t\right)^{-1} \left(\frac{d\widehat{v}}{dt} - \mathbf{A}\left(\widehat{v},t\right)\right). \qquad (7.44)$$

Here, $\mathbf{A} = (A_1, A_2, ..., A_N)$, where A_2 would be the drift coefficient (change in time of total current) for the neuron located at $\mathbf{x}_2$. Taking, $\mathbb{B} = D\mathbb{I}$, with $\mathbb{I}$ an identity matrix and D a constant diffusion coefficient, Eq. (7.43) becomes,

$$P\left(\widehat{v},t+\tau \mid \widehat{v}_0,t\right) = \left(\frac{1}{2\pi D\tau}\right)^{3/2} \exp\left\{-\frac{1}{2D}\left[\left(\frac{\triangle\widehat{v}}{\tau}\right) - \mathbf{A}\left(\widehat{v},t\right)\right]^2 \tau\right\}. \qquad (7.45)$$

Another direction for study is to incorporate the observation that real brain tissue shows neurons embedded in a mass of glial cells such as astrocytes and oligodendrocytes. The proportion is roughly 1 neuron to 10 glial cells. Recent work has shown that such glial cells do not just play an ancillary role in the activity of neurons. Indeed, they are observed to enhance synaptic transmission of signals [160]. It would be helpful to see how this would affect modelling approaches. Moreover, it is also of interest to see how boundary conditions at cellular membranes can be handled within the framework of white noise path integrals. For example, mixed boundary conditions such as those discussed in [80] can be accommodated in the white noise path integral approach to obtain a closed form for the conditional probability density function.

7.2 Neuronal Firing Rate

Various parts of the brain are characterized by highly oscillatory activities [14] with neurons integrating all input signals before firing and sending output signals to adjacent neurons. How neurons code and decode messages, therefore, may partly reside in the firing rate of neurons which has been referred to as the rate code [131]. A particularly interesting case would be the so-called hub neuron which is a convergence point of links to many other neurons which could fire in synchrony [40]. To model firing rates, one possibility is to use Eq. (3.6) with an oscillatory memory function given by Eq. (3.17). We let $x(\tau)$ be a firing rate with an initial value x_0, and write,

$$x(\tau) = x_0 + \int_0^\tau \sin^{\frac{1}{2}}(\tau - t)\sqrt{J_0(t)}\omega(t)\,dt, \qquad (7.46)$$

where $J_\nu(t)$ is the Bessel function. Equation (7.46) clearly shows that the firing rate at time τ is a result of integrating past events and representations of presynaptic inputs at prior times $t < \tau$. Since the white noise variable $\omega(t)$ may be positive or negative, Eq. (7.46) could reflect the interplay of excitatory and inhibitory signals that affect the rate $x(\tau)$. Experimentally, the measurement of firing rates is done by averaging the spike count within a certain time window [131; 59]. In our approach, the probability that the firing rate is $x(T) = x_T$ at time $\tau = T$, if it started at x_0, would then be described by the probability density function Eq. (3.18). An analytical model, however, should be closely aligned with the experimentally accessible firing rate of a real neuron. Nonetheless, this example illustrates how other types of memory functions may be employed in investigating rate codes of neurons using the white noise variable $\omega(t)$.

7.3   Interneuronal Connections with Memory

In this section, we present fractional stochastic path integral approach to derive the conditional probability density function (pdf) for the propagation of neuronal signals through interconnected chains of neurons. A number of considerations make this approach interesting as a source of insight in the investigation of current core problems of neuroscience, i.e., connectivity, memory, correlations and synchrony. In particular, we recall some salient properties of these biological systems: (i) Populations of neurons are typically extremely dense, with dendritic arborizations and synaptic branching linking a single neuron to around 10,000 other neurons [38; 42]. (ii) An enormous number of possible link configurations for the synaptic transmission of signals between initiator and distant target receptor neurons, with an individual neuron typically also receiving signals from other neurons that are not part of the chain directly linking initiator and final receiver neurons. Indeed, a diversity of input signals are integrated by a neuron for each output signal it sends out to adjacent neurons in a chain. (iii) Interneuronal connectivity is determined by synaptic activity that is highly variable and dynamic. At the cellular level, various experiments, indicate that memory is encoded by changes in the synaptic strength or connectivity between communicating neuronal cells in the brain [128; 64; 149; 183]. These considerations lead us to search for an analytical framework whose elements could be matched *vis-á-vis* biological features. First, we use the Feynman path integral approach since it precisely works for systems with large numbers of degrees of freedom, having been conceptually designed to work as a 'summation over all possible histories' of a physical system. Second, a stochastic description appears particularly suitable since lines of communication within networks of neurons are not deterministic in the sense of well-defined and predictable trajectories for signal transmission.

We are thus led to the sum over all network paths as the Hida-Streit generalized stochastic functional integral over the infinite-dimensional white noise space of distributions [106; 196]. A distinct characteristic of the path integral approach would be its being global in nature, in contrast to other network and graph-theoretical approaches that work bottom-up starting with the simplest point-to-point connections, and inductively going up to more connections and branches. Clearly, the latter would be bound to encounter increasing demand for computational time and power to comprehensively handle increasingly complex neuronal architecture in contrast to the former approach. The global nature of the integral also makes it convenient for handling boundary conditions and effects of topological constraints [21]. Finally, the propagation of signals or points in the interneuronal flow of information can be endowed with memory, with each point along a signal transmission trajectory depending on an earlier neuronal configuration. Thus, as a practical example, we parametrize fluctuations in the path integral in terms of fractional Brownian motion (fBm) characterized by the Hurst index H. An advantage of fBm is its being essentially non-Markovian, having long-range spatial and temporal correlations [176]. The mathematical foundations of fBm have been developed to a high level which allows easier application to physical and biological systems [148; 155; 156].

The treatment here differs from earlier works with stochastic analysis or path integrals, e.g., (i) random walk models for the activity of single neurons [92], (ii) stochastic jump processes for the dynamics of neuronal membrane depolarization potentials [39], (iii) path integral framework for evaluating the partition function for whole neural network dynamics [13], (iv) path integral evaluation of the potential at different points along dendritic trees [1], (v) an algorithm for evolution of short-term memory in the statistical mechanics of neocortical interactions [111], (vi) path integral for the propagation of cerebellar signals [187], and (vii) spin coherent state path integrals for the neural network master equation describing neural activity [166]. Finally, the way we implement fBm parametrization differs from earlier works on the fractal features of neuronal structures and behavior as reviewed in [209].

In the following, we set up the fractional path integral for the propagation of signals between neurons, followed by its analytical evaluation for the pdf. This is followed by a discussion of information on neuronal connectivity and correlations that can be derived from the pdf. We can then look at how these properties could change as one studies connectivity in a local cluster, modelling, for example, one cubic millimeter of brain tissue. We could also draw insights as to how to extrapolate to macroscales where neuronal connectivity can be in different anatomical areas [206]. This is important since, as pointed out in Reference [131], a transmission line linking single neurons may be unreliable.

7.3.1 *Parametrization of Neuronal Connections*

In view of the high density of neuronal connections, a large number of possible neuronal communication chains may conduct signals starting from an initiator neuron at $\mathbf{r}_0$, through intermediate neurons with their corresponding axons, synapses and dendrites, and finally received by a target receptor neuron at $\mathbf{r}$. One could then visualize a random distribution of signal transmission trajectories or diffusion paths starting at $\mathbf{r}_0$ and ending at $\mathbf{r}$. Figure 7.1 shows two possible communication trajectories. Note that, for biological systems, neurons which are part of the communication chain may also receive signals from other neurons that are not part of the chain connecting $\mathbf{r}_0$ and $\mathbf{r}$. Here, however, we are considering only the transmission of actual output signals. Any variations in the information would be carried by the fBm parametrization. Thus, we have for the diffusion paths,

$$\mathbf{r}(s) = \mathbf{r}_0 + \mathbf{B}^H(s) , \qquad (7.47)$$

where we note that, the parameter s, $0 \leq s \leq L$, is a segmental path length which increases at each segment or step of the fBm and L is a total path length for the connections between $\mathbf{r}(0)$ and $\mathbf{r}(L)$, $\mathbf{r} = (x, y, z)$. This parametrization, in lieu of time t, is allowed by defining short path intervals $\Delta s \approx \mathrm{v}t$, where v is an average velocity of Brownian motion. This is in analogy to the path integral parametrization of polymer chains in terms

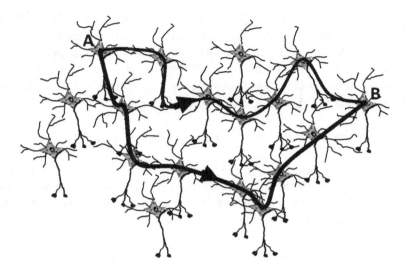

Fig. 7.1. There are numerous possible paths linking communicating neurons such as the top and bottom paths linking neurons A and B.

of monomer lengths by Edwards [73]. The $\mathbf{B}^H = \left(B_x^H, B_y^H, B_z^H\right)$ with components B_i^H $(i = x, y, z)$, is a fBm in the Riemann-Liouville fractional integral representation defined by [156],

$$B_i^H(L) = \frac{1}{\Gamma\left(H + \frac{1}{2}\right)} \int\limits_0^L (L - s)^{H - \frac{1}{2}} \, dB_i\left(s\right) . \qquad (7.48)$$

Again we note that the derivative of the Brownian motion, $\omega_i = dB_i/ds$, which is white noise, may also be used to write $dB_i = \omega_i ds$ in Eq. (7.48), and H is the Hurst index.

7.3.2 *Sum-over-all-paths between Communicating Neurons*

We now consider the sum-over-all possible connections or configurations of paths $\mathbf{r}(s)$ $(0 \leq s \leq L)$ that start with an initiator neuron located at $\mathbf{r}(0) = \mathbf{r}_0$ and end at the target receptor neuron located at $\mathbf{r}(L) = \mathbf{r}_L$. Specifically, we allow all possible paths described by Eq. (7.47) subject to the constraint of terminating at $\mathbf{r}_L$. In other words, regardless of their configuration, contributing paths are only those that satisfy the constraint,

$\delta\left(\mathbf{r}\left(s\right)-\mathbf{r}_{L}\right)$. What then is the distribution probability for all these connecting paths which satisfy the constraint? The conditional probability density function $P\left(\mathbf{r}_{L},\mathbf{r}_{0};L\right)$, gives the probability that the position $\mathbf{r}(s)$ is $\mathbf{r}_{L}$ at $s=L$, given the initial condition that its value is $\mathbf{r}_{0}$ at $s=0$. Its value can be obtained by evaluating the expectation value of the delta function constraint, i.e.,

$$P\left(\mathbf{r}_{L},\mathbf{r}_{0};L\right)=E\left(\delta\left(\mathbf{r}\left(s\right)-\mathbf{r}_{L}\right)\right)=\int\delta\left(\mathbf{r}\left(s\right)-\mathbf{r}_{L}\right)d\boldsymbol{\mu}\,,\qquad(7.49)$$

where $d\boldsymbol{\mu}=(d\mu_{x},d\mu_{y},d\mu_{z})$ is a Gaussian white noise measure defined by the characteristic functional, Eq. (2.6) [106]. The probability density is then expressed as a weighted average over fBm paths with end-point pinning. Considering the x degree of freedom, for simplicity, we have,

$$P\left(x_{L},x_{0}\right)=\int\delta\left(x_{0}+B_{x}^{H}-x_{L}\right)d\mu_{x}\,.\qquad(7.50)$$

The delta function can be written in terms of its Fourier representation, thus,

$$P\left(x_{L},x_{0}\right)=\int\left(\frac{1}{2\pi}\int\limits_{-\infty}^{+\infty}\exp\left\{ik\left[\left(x_{0}-x_{L}+B_{x}^{H}\right)\right]\right\}dk\right)d\mu_{x}$$

$$=\frac{1}{2\pi}\int\limits_{-\infty}^{+\infty}dk\exp\left\{ik\left[\left(x_{0}-x_{L}\right)\right]\right\}$$

$$\times\int\exp\left\{ik\int\limits_{0}^{L}\frac{\left(L-s\right)^{H-\frac{1}{2}}\omega_{x}ds}{\Gamma\left(H+\frac{1}{2}\right)}\right\}d\mu_{x}\,.\qquad(7.51)$$

The integration over ω_{x} then appears as a Fourier transform of $d\mu_{x}$, (see Eq. (2.6)) i.e.,

$$\int\exp\left\{ik\int\limits_{0}^{L}\frac{\left(L-s\right)^{H-\frac{1}{2}}\omega_{x}ds}{\Gamma\left(H+\frac{1}{2}\right)}\right\}d\mu_{x}=\int\exp\left\{i\int\limits_{0}^{L}f\left(s\right)\omega_{x}\left(s\right)ds\right\}d\mu_{x}$$

$$=\exp\left\{-\frac{1}{2}\int\limits_{0}^{L}f\left(s\right)^{2}ds\right\}\,,\qquad(7.52)$$

where the right-hand side is just the characteristic functional [106] and

$$f(s) = \frac{k(L-s)^{H-\frac{1}{2}}}{\Gamma\left(H+\frac{1}{2}\right)}.$$ (7.53)

Integrating $\int f(s)^2 ds$ in Eq. (7.52) and using this in Eq. (7.51) we have,

$$P(x_L, x_0) = \frac{1}{2\pi} \int\limits_{-\infty}^{+\infty} \exp\{ik(x_0 - x_L)\} \exp\left\{-\frac{k^2 L^{2H}}{4H\,\Gamma^2\left(H+\frac{1}{2}\right)}\right\} dk.$$ (7.54)

The remaining Gaussian integral can be evaluated to yield,

$$P(x_L, x_0) = \sqrt{\frac{H\,\Gamma^2\left(H+\frac{1}{2}\right)}{\pi L^{2H}}} \exp\left\{-\frac{H\,\Gamma^2\left(H+\frac{1}{2}\right)(x_L - x_0)^2}{L^{2H}}\right\}.$$ (7.55)

The y and z degrees of freedom can be treated in the same manner to yield $P(y_L, y_0)$ and $P(z_L, z_0)$ with a similar form as that of $P(x_L, x_0)$. In the following section, it would be sufficient to consider the properties of $P(x_L, x_0)$.

Note that the probability distribution function $P(x_L, x_0)$, with $0 \le s \le L$, obeys an effective Fokker-Planck equation of the form [46],

$$\frac{\partial}{\partial s}P(x, s) = \frac{s^{2H-1}}{2\Gamma^2\left(H+\frac{1}{2}\right)}\frac{\partial^2}{\partial x^2}P(x, s).$$ (7.56)

When the Hurst index is $H = 1/2$, Eq. (7.56) reduces to the usual diffusion equation for Brownian motion.

7.3.3 *Distance and Connectivity between Neurons*

What information can be extracted from the probability distribution function $P(x_L, x_0)$? If we designate, $\mathbf{x} = x_L - x_0$, as the physical distance along the x-axis between the neurons at $\mathbf{r}_0$ and $\mathbf{r}_L$, we can graph the probability distribution function, Eq. (7.55), as a function of distance $\mathbf{x}$. In particular, we can do this for three different values of the Hurst index (see Figure 7.2). In the figure, the solid line is for $H = 0.5$ (ordinary Brownian motion), while the dashed line, $H = 0.3$, represents connections modelled by a suppressed diffusion. The dotted line, $H = 0.8$, describes paths for the case of enhanced diffusion. All three graphs are for the same total path length L.

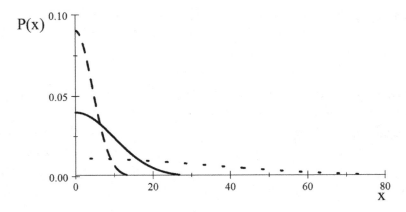

Fig. 7.2. Solid line: $H = 0.5$ (ordinary Brownian motion); Dash: $H = 0.3$; Dots: $H = 0.8$. The probability distribution function shows that two neurons are more likely to be far apart when $0.5 < H < 1$.

As exemplified by the curves in Figure 7.2, if two neurons that are near each other (i.e., small values of $x = x_L - x_0$) communicate through neuronal chains of path length L, the connections involved would more likely correspond to the sub-diffusion type ($0 < H < 0.5$). On the other hand, two communicating neurons far from each other (i.e., large x), such as those located at different regions of the brain would likely have connection patterns corresponding to an enhanced diffusion process ($0.5 < H < 1$).

Equation (7.55) therefore indicates that connectivity could be scale-dependent where nearby neurons (described by $0 < H < 0.5$) and neurons significantly separated (e.g., for $0.5 < H < 1$) have different connectivity patterns characterized by the Hurst index H. This reminds us of a recent experiment with multiple-tetrode recordings from primary visual cortex of macaque monkeys which showed a scale-dependent structure of cortical interactions [165]. The results showing how connectivity is organized in the cortex at multiple scales were based on recordings where a tetrode isolates several neurons within a radius of $\sim 150\ \mu\text{m}$, monitoring small groups of 3 to 6 neurons, as well as tetrodes separated by distances ranging from 600 μm to several mm. The investigators pointed out a curious result: unlike neurons separated by large distances, local clusters of neurons exhibited flexible correlations that are rapidly reorganized by visual input. In our framework, local clusters would correspond to short memory processes where $0 < H < 0.5$.

7.3.4   *Interneuronal Correlation and Synchrony*

The correlation of spiking neurons, both at the temporal and spatial scales, has been the subject of many experimental investigations. The spatial properties and distance between correlated neurons, however, appear to be an important factor in understanding the mechanism for correlation. Recent technological advances in using multielectrode arrays [190] and calcium imaging techniques [57] have allowed measurement of correlation between activities of neuron pairs *in vivo*. With multiple tetrode recordings [165], for instance, firing patterns can be compared among local clusters of neurons (< 300 μm apart) with those of neurons separated by larger distances (600–2500 μm). Results show that firing patterns at larger distances are predicted by pairwise interactions, while patterns within local clusters show evidence of high-order correlations. Correlations are taken to manifest primarily in synchronized firing of neurons, that is when two or more spiking neurons fire jointly [131]. It is thus of interest to study both short-range and long-range communications between neurons separated by some distance or by layers of intermediate neurons. Furthermore, synchrony was found evident between nearby neurons separated by 3 mm or less, while correlated variability decreased slowly with distance but is still significant between neurons separated by 10 millimeters.

In the present formalism where we idealize the connections as fBm paths, the correlation between two neurons at points along the chain can be calculated. In particular, the expectation value or correlation is of the form [156],

$$E\left(x\left(s_1\right) x\left(s_2\right)\right) \simeq \frac{1}{2}\left(\left|s_1\right|^{2H} + \left|s_2\right|^{2H} - \left|s_1 - s_2\right|^{2H}\right), \qquad (7.57)$$

where s_1 and s_2 are distance parameters specifying the locations of two correlated neurons along the contour path length (see Figure 7.3). Equation (7.57) tells us that correlation largely depends on distances of neurons along the path length. In particular, for two neighboring neurons where $s_1 \approx s_2$, the negative term, $\left|s_1 - s_2\right|^{2H}$ is negligible resulting in a relatively higher correlation. On the other hand, if two neurons have a large separation along the path length, i.e., $\left|s_1 - s_2\right|^{2H}$ becomes large in Eq. (7.57), the correlation between neurons could dramatically decrease. This could give insight on the interpretation of results of experiments showing that correlated neuronal activities depend on the scale. For instance, sampling of cortical activities at different scales showed firing patterns for local clusters of neurons (< 300 μm apart) with high-order correlations, which differed

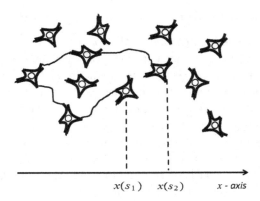

Fig. 7.3.   Two neurons at $x(s_1)$ and $x(s_2)$, which could be physically near each other, may communicate using a much longer path length L.

from neurons separated by larger distances (600–2500 μm) which can be predicted by pairwise interactions [165].

   Equation (7.57) also depends on the value of the Hurst index H which indicates whether the connection stucture is of the suppressed or enhanced type of diffusion. In particular, as observed in the previous section, two communicating neurons which are physically near each other (small values of $\mathbf{x}$, but may have large contour separation $|s_1 - s_2|$), would have connections that are more likely to be sub-diffusive.

   The mean squared distance between two points of the neuronal chain can likewise be computed and is given by,

$$E\left([x(s_1) - x(s_2)]^2\right) = |s_1 - s_2|^{2H}. \qquad (7.58)$$

Again, we have to distinguish between the actual physical distance between two neurons given by, $x(s_1) - x(s_2)$ along the x-axis, and their separation along the contour path given by the length, $|s_1 - s_2|$. Two neighboring neurons $(x(s_1) \approx x(s_2))$ may have a large separation, $|s_1 - s_2|$, along the path length (see Figure 7.3).

   Equation (7.58) can further be generalized for higher moments, $n \geq 1$, to yield the result [156],

$$E\left([x(s_1) - x(s_2)]^n\right) = \frac{2^{\frac{n}{2}}}{\sqrt{\pi}}\Gamma\left(\frac{n+1}{2}\right)|s_1 - s_2|^{nH}. \qquad (7.59)$$

The expression, Eq. (7.59), tells us that even if two neurons are physically very close to each other, i.e., $x(s_1) \approx x(s_2)$, the expectation value of $[x(s_1) - x(s_2)]^n$ could be large if their separation along the contour length, $|s_1 - s_2|$, is large. Equation (7.59) may be checked with empirical investigations of paired neurons and higher-order correlations.

7.4 Scaling Property for Neuronal Clusters

Among the challenges facing theoretical and experimental investigations in neuroscience is the tracking of activities between populations of neurons. Here one essentially goes from studying small groups of connected neurons, to bigger clusters or assemblies of neurons, all the way up to connected large neuron populations. The connectivity to be investigated this time involves a neuron population receiving inputs from a previous neuron population as it projects to a subsequent neuron population (see Figure 7.4).

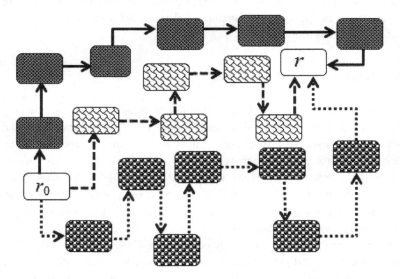

Fig. 7.4.   Each rectangle represents a neuron population. There are many paths for signals to go from r_0 to another neuron population at r.

This is closer to biologically realistic models with relationships between local activities of neurons linked by synapses to large-scale global connectivity involving networks of brain regions connected by anatomical tracts. The inherent self-similarity property of fBm allows the present formulation to offer a way of tracking small-scale neuronal activities to large-scale activities between neuron populations. One could implement a description of neuronal connectivity by introducing a scale factor $b > 0$ such that the path length s and distance along x transform as,

$$s \to bs \quad ; \quad x \to b^H x. \tag{7.60}$$

It is easy to check that under the transformations given by Eq. (7.60), the probability distribution function Eq. (7.55) transforms accordingly as,

$$P\left(b^H x_s, b^H x_0; bs\right) = b^{-H} P\left(x_s, x_0; s\right). \tag{7.61}$$

Equation (7.61) states that the probability distribution function for connectivities on the large scale described by $P\left(b^H x_s, b^H x_0; bs\right)$ has the same property and behavior as the small-scale probability distribution function $P\left(x_s, x_0; s\right)$ except for a factor of b^{-H}. Scale transformations may describe connectivity between neuron populations. Although anatomically segregated regions of the brain are highly specialized in their functions, a self-similarity property may, perhaps, help characterize the structure and function of the brain.

7.5 Synchronous Burst Rates

Another approach in investigating the neural code is looking at the simultaneous joint firing of several or more neurons. This occurs when two or more neurons are correlated. Such synchronous firing means that different neurons generate at least one spike within the same time window which constitutes a new signal. Neural information stored in this simultaneous burst of correlated neurons is referred to as a temporal code [131]. Studies on synchronous burst firing of neurons are important, for example, in understanding epilepsy [162; 193].

Neurons that synchronize and coordinate their spikes to encode one signal operate at a different timescale compared to firing rates of an individual neuron [131]. Measurements of correlated firing would be ≤ 10 ms, while individual neuronal spike counts usually involve time resolutions of ≥ 100 ms. Since different neurons now collaborate to encode one signal, a memory function quite different from that of individual neurons may be needed. Designating $x\left(\tau\right)$ as the rate of sychronous firing of neurons, a candidate is the exponentially modified white noise which gives,

$$x\left(\tau\right) = x_0 + \int\limits_0^\tau \frac{\left(\tau - t\right)^{(\mu-1)/2} e^{-\beta/2t}}{t^{(\mu+1)/2}} \omega\left(t\right) dt. \tag{7.62}$$

In Eq. (7.62), the exponential factor, $e^{-\beta/2t}$ dominates only for short time scales which could be typical of bursting events. As in the previous example, the positive or negative sign of the white noise variable $\omega(t)$ could represent competing excitatory and inhibitory signals influencing burst rates. Equation (7.62) also allows differentiation between a short-memory domain for $0 < \mu < 1$, and long-memory for $1 < \mu$. These domains may prove useful in understanding epileptic seizures where a distinction in synchronization is made: (a) between local and distant cortical areas [74]; and (b) at the initiation phase (characterized by low frequency ~ 35 Hz synchrony) and termination phase (high frequency ~ 120 Hz synchrony) of seizures [191]. The probability that the burst rate would be $x(T)$ at time $\tau = T$ if it started at x_0 would be given by the probability density function Eq. (3.21). The constant β in this case is an adjustable parameter for possible matching with actual experiments. Other forms of memory function can, of course, be used for modelling purposes.

Chapter 8

Biopolymers

The thousands of proteins in the human body can only function biologically if each attains a specific three-dimensional structure or conformation. The mechanisms involved in the attainment of a specific conformation, however, is far from understood. There is, in fact, an interdisciplinary effort to solve this protein folding problem. The application of computational methods, for instance, has shown successes in simulating protein conformation [217]. Employing more realistic potentials, or an all-atom simulation for the folding process, however, is constrained by computing time and resources which often limit the study to short polypeptide chains [213]. It might be helpful then to explore various analytical methods to understand protein folding.

Since we will be looking at proteins diffusing through a crowded aqueous cell to attain their native structure, we here consider the Fokker-Planck equation. In particular, we illustrate how white noise path integration facilitates the solution of the Fokker-Planck equation. Since we are now dealing with Markovian processes, the path parametrization Eq. (1.13) for this chapter is the form of $f(\tau - t)$ given by Eq. (3.14) where, specifically, the value of the Hurst index $H = \frac{1}{2}$ is chosen. This effectively gives $f(\tau - t) = 1$, where memory of past events is erased. Our discussion will then focus on the formation of the α-helical segment which is an ubiquitous secondary structure in proteins.

8.1 Modelling Diffusive Polypeptides

The solution of the Fokker-Planck equation for small intervals of time, $\tau \ll 1$, is given by the conditional probability density [29; 181]

$$P(\mathbf{r}, t + \tau \mid \mathbf{r}_0, t) = \left(\frac{1}{2\pi D\tau}\right)^{3/2} \exp\left\{-\frac{1}{2D}\left[\left(\frac{\triangle \mathbf{r}}{\tau}\right) - \mathbf{A}\right]^2 \tau\right\}, \quad (8.1)$$

where $\mathbf{A}\left(\mathbf{r}_0, t\right)$ is the drift vector, and D a constant diffusion coefficient. With a view to investigating polymer chains, we denote the length travelled in time $\tau \ll 1$ as, $\triangle s = v\tau$, where v is the average speed of the random motion, and Eq. (8.1) acquires the form,

$$P\left(\mathbf{r}, s_0 + \triangle s \mid \mathbf{r}_0, s_0\right) = \left(\frac{v}{2\pi D \triangle s}\right)^{3/2} \exp\left\{-\frac{v}{2D}\left[\left(\frac{\triangle \mathbf{r}}{\triangle s}\right) - \frac{\mathbf{A}}{v}\right]^2 \triangle s\right\},$$

(8.2)

with $\triangle s \ll 1$. Application of the Chapman-Kolmogorov equation leads to a conditional probability density given by the path integral (see Section 6.3),

$$P\left(\mathbf{r}_1, L \mid \mathbf{r}_0, 0\right) = \int \exp\left\{-\frac{3}{2l}\int_0^L \left[\frac{d\mathbf{r}}{ds} - \frac{l}{3D}\mathbf{A}\right]^2 ds\right\} \mathcal{D}\left[\mathbf{r}\right],$$

(8.3)

where $l = 3D/v$, which has a dimension of length, L is the total path length, and the integral is taken over all paths $\mathbf{r}(s)$ starting at $\mathbf{r}(0) = \mathbf{r}_0$ and ending at $\mathbf{r}(L) = \mathbf{r}_1$, with $0 \leq s \leq L$.

The path integral, Eq. (8.3), can be used to model a polypeptide chain where various diffusion paths with endpoints at $\mathbf{r}_0$ and $\mathbf{r}_1$ are viewed as possible conformations of a biopolymer with the same endpoints. Here, N is the number of amino acids forming the chainlike protein with each monomer being of length l. In modelling the biopolymer, trajectories in the path integral would consist of N steps each of length l, such that $L = Nl$ represents the length of the polymer [73]. The paths in Eq. (8.3) can be parametrized in terms of ordinary Brownian motion,

$$\mathbf{r}(s) = \mathbf{r}_0 + \kappa \mathbf{B}\left(s\right),$$

(8.4)

where $\mathbf{r} = (x, y, z)$, $\mathbf{r}_0$ is the initial point, and $\mathbf{B}\left(s\right)$ are Brownian fluctuations with κ a constant. The derivative of the path $\mathbf{r}(s)$ is given by $d\mathbf{r}/ds = \kappa\,\boldsymbol{\omega}\left(s\right)$, where $\boldsymbol{\omega}\left(s\right) = d\mathbf{B}/ds$ is a Gaussian random white noise variable [106]. The parametrization Eq. (8.4), therefore, once incorporated into Eq. (8.3) leads to an integrand which is a functional $\Phi\left[\boldsymbol{\omega}\left(s\right)\right]$. Its evaluation makes use of the Hida-Streit method [106; 196] for the quantum mechanical propagator.

8.2 Helical Conformations

Since in the study of proteins the aim is to investigate helical structures, it is convenient to use circular cylindrical coordinates $\mathbf{r} = (\rho, \varphi, z)$ where helical or winding conformations can be viewed as windings about the z-axis. A simplification can be invoked knowing that actual α-helical structures of proteins have a typical radius of $R = 0.25$ nm. We can, therefore, set the radial variable as $\rho = R$, and allow the φ variable to track the clockwise or anti-clockwise windings around the z-axis. Pictorially, we would be dealing with Brownian paths on the surface of a cylinder of radius R, modulated by a drift coefficient. With a helical biopolymer oriented along the z-axis, we can take advantage of its axial symmetry by projecting the winding behavior of a biopolymer onto the $(\rho - \varphi)$ plane [23; 27]. This leaves us the φ-degree of freedom to investigate winding probabilities, and we can take $\mathbf{A} = (0, h(s), 0)$ where $h(s)$ is the φ-component of the drift vector.

Consider then a polymer chain lying on the plane perpendicular to the z-axis a distance $\rho = R$ from the origin. Given the endpoints at φ_0 and φ_1 we have, corresponding to Eq. (8.3), the conditional probability density,

$$P(\varphi_1, \varphi_0) = \int \exp\left[-\frac{1}{l}\int_0^L \left[R\left(\frac{d\varphi}{ds}\right) - \frac{l}{2D}h(s)\right]^2 ds\right] \mathcal{D}[\varphi]. \qquad (8.5)$$

In an aqueous environment, the 20 different amino acids forming the protein can be hydrophilic or hydrophobic, polar or non-polar with varying hydropathy indices, so we allow the value of the drift coefficient $h(s)$, with $0 \le s \le L$, to also vary at each length segment along the chainlike molecule. Corresponding to Eq. (8.4), the variable φ can be parametrized as,

$$\varphi(s) = \varphi_0 + \left(\sqrt{l}/R\right) B(s) \qquad (8.6)$$

where φ_0 is the starting point and B the Brownian fluctuation.

8.3 Path Integral for Helical Conformations

We shall now illustrate essential steps in evaluating Eq. (8.5). The parametrization Eq. (8.6) leads to an integrand in Eq. (8.5) which is a functional of the white noise variable $\omega(s)$. In particular, we have

$$\exp\left[-\frac{R^2}{l}\int_0^L \left(\frac{d\varphi}{ds}\right)^2 ds + \frac{R}{D}\int_0^L h(s)\left(\frac{d\varphi}{ds}\right) ds - \frac{l}{4D^2}\int_0^L h(s)^2 ds\right]$$

$$= \exp\left[-\int_0^L \omega(s)^2 ds + \frac{\sqrt{l}}{D}\int_0^L h(s)\ \omega(s) ds - \frac{l}{4D^2}\int_0^L h(s)^2 ds\right]. \qquad (8.7)$$

With an integrand being a white noise functional, the path integration over $\mathcal{D}[\varphi] = d^\infty\varphi$ becomes an integral over the Gaussian white noise measure $d\mu(\omega)$. Using Eq. (2.7), the integral over $\mathcal{D}[\varphi]$ becomes an integral over $N_\omega \, d^\infty\omega = \exp[\frac{1}{2} \int \omega(s)^2 \, ds] \, d\mu(\omega)$. From the factor $\exp[\frac{1}{2} \int \omega(s)^2 \, ds]$, and the exponential, $\exp\{-\int \omega(s)^2 ds\}$ on the right-hand side of Eq. (8.7), we define the Gauss kernel I_0 given by,

$$I_0 = N \exp\left(-\frac{1}{2}\int\limits_0^L \omega(s)^2 \, ds\right), \tag{8.8}$$

where N is an appropriate normalization factor.

The path parametrization, Eq. (8.6) fixes the initial point φ_0, but the endpoint φ_1 of the Brownian paths still needs to be pinned down. To do this, we use the Donsker delta function discussed in Section 2.4, $\delta(\varphi(L) - \varphi_1)$. Furthermore, Eq. (8.5) deals with a Brownian motion confined to a circular topology. A more detailed presentation of such a topologically constrained problem is found in Chapter 10. In the meantime, we simply take the Brownian paths to be classified topologically depending on the direction, clockwise or counterclockwise, and the number of times the polymer chain turns around the origin of the plane before ending at φ_1. These paths can be characterized by winding numbers [185] $n = 0, \pm 1, \pm 2, ...$, where $n > 0$ signifies n turns counterclockwise around the origin; $n < 0$ means $|n|$ turns clockwise, and $n = 0$ signifies no winding[1]. To incorporate these possibilities, we use $\delta(\varphi(L) - \varphi_1 + 2\pi n)$, to reflect the topologically inequivalent trajectories that end at φ_1. This delta function summed over all possible n, together with Eqs. (8.7) and (8.8), allows us to write the probability function, Eq. (8.5) as,

$$P(\varphi_1, \varphi_0) = \sum_{n=-\infty}^{n=+\infty} \int I_0 \exp\left[\frac{\sqrt{l}}{D}\int\limits_0^L h(s)\,\omega(s)ds - \frac{l}{4D^2}\int\limits_0^L h(s)^2 \, ds\right]$$

$$\times \delta(\varphi(L) - \varphi_1 + 2\pi n)\, d\mu(\omega)\,. \tag{8.9}$$

The evaluation of Eq. (8.9) is facilitated by writing the Fourier represen-

---

[1] A path moving clockwise, but no winding may be labelled by $n = 0^-$, and a path going counterclockwise with no winding by, $n = 0^+$.

tation of the delta function, i.e.,

$$P(\varphi_1, \varphi_0) = \frac{1}{2\pi} \exp\left(-\frac{l}{4D^2} \int_0^L h(s)^2 \, ds\right)$$

$$\times \sum_{n=-\infty}^{n=+\infty} \int d\lambda \exp\left[i\lambda \left(\varphi_0 - \varphi_1 + 2\pi n\right)\right]$$

$$\times \int I_0 \exp\left\{i \int_0^L \left[-\frac{i\sqrt{l}}{D} h(s) + \lambda \frac{\sqrt{l}}{R}\right] \omega(s) ds\right\} d\mu(\omega).$$

(8.10)

The integration over $d\mu(\omega)$ is just the T-transform of the Gauss kernel I_0, Eq. (2.27), which yields,

$$T(I_0)(\xi) = \int I_0 \exp\left[i \int_0^L \omega(s)\xi(s) \, ds\right] d\mu(\omega)$$

$$= \exp\left[-\frac{1}{4} \int_0^L \xi(s)^2 \, ds\right],$$

(8.11)

where $\xi(s) = \sqrt{l}\left[(-i/D) h(s) + (\lambda/R)\right]$. Thus, Eq. (8.10) acquires the form,

$$P(\varphi_1, \varphi_0) = \frac{1}{2\pi} \sum_{n=-\infty}^{n=+\infty} \int d\lambda \exp\left(-\frac{lL}{4R^2}\lambda^2\right)$$

$$\times \exp\left\{i\left[(\varphi_0 - \varphi_1 + 2\pi n) + \frac{l}{2DR} \int_0^L h(s) \, ds\right]\lambda\right\}.$$

(8.12)

The remaining integral over λ may be evaluated in two ways. Employing the Poisson sum formula,

$$\frac{1}{2\pi} \sum_{n=-\infty}^{+\infty} \exp(in\kappa) = \sum_{m=-\infty}^{+\infty} \delta(\kappa + 2\pi m),$$

(8.13)

Eq. (8.12) becomes,

$$P(\varphi_1, \varphi_0) = \frac{1}{2\pi} \sum_{m=-\infty}^{+\infty} \int \exp\left(-\frac{lL}{4R^2}\lambda^2\right)$$

$$\times \exp\left\{i\left[(\varphi_0 - \varphi_1) + \frac{l}{2DR}\int_0^L h(s)\,ds\right]\lambda\right\}$$

$$\times \delta(\lambda + m)\,d\lambda, \tag{8.14}$$

and the integration is facilitated by the delta function to yield the result,

$$P(\varphi_1, \varphi_0) = \frac{1}{2\pi} \sum_{m=-\infty}^{+\infty} \exp\left(-\frac{lL}{4R^2}m^2\right)$$

$$\times \exp\left\{-im\left[(\varphi_0 - \varphi_1) + \frac{l}{2DR}\int_0^L h(s)\,ds\right]\right\}$$

$$= \frac{1}{2\pi}\boldsymbol{\theta}_3(u). \tag{8.15}$$

Here, the $\boldsymbol{\theta}_3(u)$ is the theta function [98],

$$\boldsymbol{\theta}_3(u) = 1 + 2\sum_{m=1}^{+\infty} q^{m^2}\cos(2mu), \tag{8.16}$$

where $u = -\frac{1}{2}\left[(\varphi_0 - \varphi_1) + (l/2DR)\int h\,ds\right]$ and $q = \exp\left(-Ll/4R^2\right)$. Alternatively, from Eq. (8.12), the integral over λ is evaluated as a Gaussian integral to yield,

$$P(\varphi_1, \varphi_0) = \sum_{n=-\infty}^{+\infty} P_n, \tag{8.17}$$

where

$$P_n = \sqrt{\frac{R^2}{\pi l L}}\exp\left[-\frac{R^2}{lL}\left(\varphi_0 - \varphi_1 + 2\pi n + \frac{l}{2DR}\int_0^L h(s)\,ds\right)^2\right]. \tag{8.18}$$

The P_n, Eq. (8.18), is the probability function for an n-times winding around the z-axis.

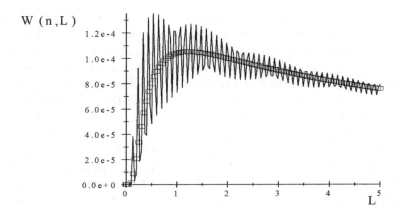

Fig. 8.1. Length-dependent winding probability for $R = 1$ nm, $l = 0.34$ nm, and $n = 2300$: solid line, $h(s) = \cos(\nu s)$, with $\nu = 46,000$; open squares, $h(s) = 0$.

The probability that a polymer chain winds n-times is given by, $W(n, L) = P_n/P(\varphi_1, \varphi_0)$. Using Eqs. (8.15) and (8.18) this yields [22; 23; 27; 28],

$$W(n, L) = R\sqrt{\frac{4\pi}{Ll}} \frac{\exp\left[-\frac{R^2}{Ll}\left(\varphi_0 - \varphi_1 + 2\pi n + \frac{l}{2DR}\int_0^L h\,ds\right)^2\right]}{\theta_3\left(\frac{(\varphi_0 - \varphi_1)}{2} + \frac{l}{4DR}\int_0^L h\,ds\right)}. \quad (8.19)$$

Equation (8.19) is an exact result obtained by evaluating Eq. (8.5). The interaction of each amino acid with the aqueous environment as well as with other monomers would be reflected in the drift coefficient $h(s)$, as s ranges along the length of a biopolymer from 0 to L. The $h(s)$ in turn modulates the winding probability $W(n, L)$ that describes a specific winding conformation. Here, a biopolymer winding counterclockwise ($n > 0$) is designated as left-handed, and the one winding clockwise ($n < 0$) as right-handed. Since the endpoints are arbitrary, we may take $\varphi_0 = \varphi_1$.

Two remarks may be noted. First the handedness, or chirality, of a polymer is made manifest in Eq. (8.19) if the drift coefficient is nonzero, i.e., $h(s) \neq 0$. Given left-handed ($n > 0$) and right-handed ($n < 0$) biopolymers, the handedness is made evident since from (8.19), one has [27; 28], $W(-n)/W(n) = \exp\left[(4\pi n R/LD)\int h\,ds\right]$. Clearly, only when $h(s) = 0$ is chiral symmetry recovered.

The second remark is on the possible effect of a self-avoiding walk (SAW) where no two monomers can occupy the same place. Recall that Eq. (8.5) is actually a projection on the $\rho - \varphi$ plane of polymer conformations in the $\mathbf{r} = (R, \varphi, z)$ space. To select those paths without self-intersection, we can use a survival probability function S_N for weighting the probability function P_n or $P(\varphi_1, \varphi_0)$ given by Eqs. (8.18) and (8.15), respectively [173]. For a large number of N steps a walk would have an average density of approximately $\zeta \simeq N/2\pi R$. The probability to survive after each step without intersection is $1 - \zeta$, and after N steps it would be $S_N \simeq (1 - \zeta)^N \simeq \exp(-N\zeta)$. For N steps with n windings, Eq. (8.18) would then be weighted by S_N as $P_n^{SAW} = P_n S_N$, and Eq. (8.15) becomes $P^{SAW}(\varphi_1, \varphi_0) = P(\varphi_1, \varphi_0) S_N$. Note that, to get the probability $W(n, L)$ that a path winds n-times, the weighting factor S_N cancels out, i.e., $W(n, L) = P_n^{SAW}/P^{SAW}(\varphi_1, \varphi_0) = P_n/P(\varphi_1, \varphi_0)$. One then essentially gets Eq. (8.19) as before.

8.4 Drift Coefficient as Modulating Function

In the present stochastic model, we have allowed the drift coefficient $h(s)$ to be length-dependent where $0 \leq s \leq L$, and L is the total length of the biopolymer. The drift coefficient $h(s)$, therefore, can carry the information of each monomer-environment interaction along the length of a biopolymer. Clearly, a choice of $h(s)$ can affect the winding probabilities as seen in Eq. (8.19). The $W(n, L)$ is an expression where linear information encoded along the length of a biopolymer can translate into a specific winding conformation for a polymer with $h(s)$ as a modulating function. The information, however, that is embodied by the possible available ways and variations of sequencing the 20 amino acids that make up the protein can be intractable. To cope with this difficulty, we expand the drift coefficient $h(s)$ in terms of a complete set of orthogonal functions. In particular, we may use, for example, the Laguerre polynomials and write,

$$h(s) = \sum_{m=0}^{\infty} k_m \alpha_m \mathbf{L}_m(\alpha_m s), \qquad (8.20)$$

where k_m and α_m are constants and $\mathbf{L}_m$ are the Laguerre polynomials. In Eq. (8.19), we have to evaluate, $\int h\, ds$, i.e.,

$$\int_0^L h\, ds = \sum_{m=0}^{\infty} k_m \alpha_m \int_0^L \mathbf{L}_m(\alpha_m s)\, ds. \qquad (8.21)$$

The integral for each Laguerre polynomial can be done [98] and we obtain,

$$\int_0^L h \, ds = \sum_{m=0}^{\infty} k_m \left[\mathbf{L}_m \left(\alpha_m L \right) - \frac{1}{m+1} \mathbf{L}_{m+1} \left(\alpha_m L \right) \right], \qquad (8.22)$$

which could be used in Eq. (8.19) to get the probability for n-times winding of the form $(\varphi_0 = \varphi_1)$,

$$W(n, L) = \left[\boldsymbol{\theta}_3 \left(\frac{l}{4DR} \sum_{m=0}^{\infty} k_m \left[\mathbf{L}_m \left(\alpha_m L \right) - \frac{1}{m+1} \mathbf{L}_{m+1} \left(\alpha_m L \right) \right] \right) \right]^{-1}$$

$$\times R \sqrt{\frac{4\pi}{Ll}} \exp \left\{ - \frac{R^2}{Ll} \left(2\pi n + \frac{l}{2DR} \right) \right.$$

$$\left. \times \sum_{m=0}^{\infty} k_m \left[\mathbf{L}_m \left(\alpha_m L \right) - \frac{\mathbf{L}_{m+1} \left(\alpha_m L \right)}{m+1} \right] \right)^2 \right\}. \qquad (8.23)$$

With Eq. (8.23) one can explore the various possible winding conformations by choosing specific values for the constants k_m and α_m. Moreover, for a very long chainlike biopolymer, $L = Nl \gg 1$, the theta function $\boldsymbol{\theta}_3(u)$ in Eq. (8.23) approaches unity. We shall adopt this for simplicity in the succeeding sections.

8.5 Helix-Turn-Helix Motif

The helix-turn-helix structural motif may be demonstrated as a special case of Eq. (8.23). Let us consider the first four terms of the series, in particular, we let $k_m = k$ for $m = 0, ..., 3$, and $k_m = 0$ for higher values of m. We obtain from Eq. (8.23) the expression,

$$W(n, L) \approx R \sqrt{\frac{4\pi}{Ll}} \exp \left\{ - \frac{R^2}{Ll} \left(2\pi n + \frac{lk}{2DR} \right. \right.$$

$$\left. \left. \times \sum_{m=0}^{3} \left[\mathbf{L}_m \left(\alpha_m L \right) - \frac{\mathbf{L}_{m+1} \left(\alpha_m L \right)}{m+1} \right] \right)^2 \right\}. \qquad (8.24)$$

Taking the values $R = 1$ nm, $l = 0.34$ nm, $\alpha_m = 1/m$, and $n = 40$, the graph of the winding probability $W(n, L)$ versus the biopolymer length L is given by Figure 8.2. The graph with circles is for $(k/D) = 10^{4.5}$, and

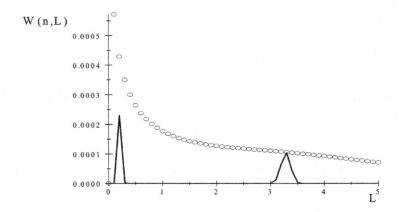

Fig. 8.2.   Graph with circles: $(k/D) = 10^{4.5}$; solid line: $(k/D) = 10^6$.

the solid line is for a much smaller value of the diffusion constant, i.e., $(k/D) = 10^6$.

The maxima of $W(n, L)$, or the peaks of the solid line in the graph of Figure 8.2, show where winding is most probable. The two peaks, however, are separated by a section where winding is strongly inhibited. This situation suggests the appearance of a helix-turn-helix structural motif in polypeptides, where a helix is likely to be formed in each of the two peaks separated by a stretched portion.

8.6   Matching with Real Proteins

It would be interesting to check if the results so far could be matched with actual protein conformation. We could take, for an arbitrary initial point, $\varphi_0 = \varphi_1$, and since for long polymers $L \gg 1$, we have $\boldsymbol{\theta}_3(u) \approx 1$. Consider a length-dependent drift coefficient given by [11],

$$h(s) = k J_{2q+1}(\nu s) , \qquad (8.25)$$

where k is a constant and $J_{2q+1}(\nu s)$ is a Bessel function of the first kind of order $2q + 1$. For the case $q = 3$, Eq. (8.19) appears as

$$W(n, L) = R\sqrt{\frac{4\pi}{Ll}} \exp\left\{ -\frac{R^2}{Ll} \left(2\pi n + \frac{lk}{2R\nu D} J'(\nu L) \right)^2 \right\} \qquad (8.26)$$

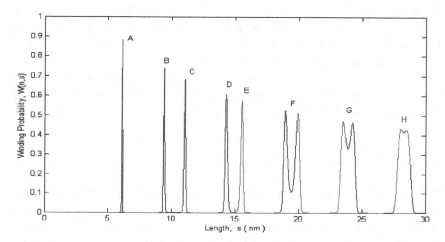

Fig. 8.3. Graph for winding number $n = -33$, radius of an α-helix, $R = 0.25$ nm, and monomer length, $l = 0.15$ nm.

using Eq. (11.1.4) of [3] where

$$J'(\nu L) = 1 - J_0(\nu L) - 2\left[J_2(\nu L) + J_4(\nu L) + J_6(\nu L)\right]. \qquad (8.27)$$

To compare with an actual protein, let us consider myoglobin which consists of 153 amino acids, 75% of which form 8-segments of α-helical conformations. Comprising about 33 turns, the α-helical segments are traditionally labelled [198] A, B,..., H, all of which are right-handed. To model myoglobin, therefore, we take $n = -33$, and use data for proteins, i.e., $R = 0.25$ nm, $l = 0.15$ nm. The resulting graph for $W(n)$ is shown in Figure 8.3 [11]. The result shows 8 major peaks which can be matched with the 8 helices of the myoglobin molecule. In between the peaks are flat regions evoking zero winding probability $W(n) = 0$ where helix formation is strongly inhibited.

As shown in the graph, helices D and E (centered at 14.3 nm and 15.5 nm, respectively) are very close to each other. This matches the experimental result that there is hardly a non-helical segment between these two helices [198]. The same holds for helices B and C (centered at 9.4 nm and 11.1 nm respectively) which are relatively close to each other compared with other helix separations. Another important point to note is that helix F, as shown, has double peaks (centered at 18.9 nm and 19.9 nm). This relates to the observation that the last part of the non-helical segment between E and F is found to form a short helix which is now called the F'-helix [205].

With this, and helices G and H (centered at 23.9 nm and 28.3 nm, respectively), Figure 8.3 apparently reproduces the information in the Protein Data Bank, which considers this molecule to have 9 instead of 8-helices.

With the huge number of proteins with known α-helical structures included in the Protein Data Bank, one can look at an inverse problem for values of parameters and the right combination of Bessel functions in Eq. (8.19) that reproduce known secondary structures.

8.7   Overwinding of DNA

Continuing experimental studies of mechanical properties of single molecules of DNA, demonstrate the counterintuitive overwinding of the strand when stretched by forces up to about 30 pN [97; 146]. Further stretching produces the expected unwinding of the coiled strand. This curious behavior has important implications for the problem of molecular recognition and conformational adjustment by proteins. However, this should fit well with other observed features of macromolecules. For example, in experiments [201; 184] to study the stepwise bending of single DNA molecules, Brownian motion was observed using video microscopy. Indeed, snapshots in time of the random-walk path travelled by the center-of-mass of a labelled λ-DNA molecule [189], validate models of polymer configurations as Brownian paths. Moreover, there is also chirality exhibited by, for example, B-form (right-handed) and its "mirror-image," Z-form (left-handed), DNA.

The sequence of bases on one DNA strand is normally matched by a complementary sequence on the other strand to form a double helix. This set-up suggests that the structure and behavior of one strand may be deduced from the other. Investigation, therefore, may focus on one nucleotide chain which winds around a second molecular chain taken to be oriented along the z-axis. We take L to be the length of the winding macromolecule composed of N repeating nucleotide units, each of length l, such that $L = Nl$. To allow for bending and twisting, each repeating unit of length l is allowed to rotate as a freely hinged rod. The nucleotide chain of our model is then viewed as a random walk consisting of N steps each of length l.

How does an extension of length affect the winding probabilities? The recent experimental result [97; 146] could serve as a guide to a choice of the modulating function $h(s)$ in Eq. (8.19) where we take $\varphi_0 = \varphi_1$.

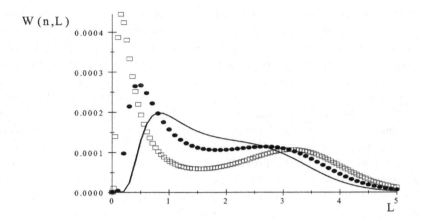

Fig. 8.4. Length-dependent winding probabilities for different values of winding n. Open Box: $n = 30$; Dots: $n = 1500$; Solid line: $n = 3000$.

An example of an integrable non-trivial modulating function is, $h(s) = k\alpha \, \mathbf{L}_2 (\alpha s)$, where $\mathbf{L}_2$ is a Laguerre polynomial with k and α constants. Using Eqs. (7.411.1), (8.970.5), and (8.970.6) of [98] one has $\int_0^L h \, ds = k \left(\frac{1}{18} \alpha^3 L^3 - \alpha L + \frac{2}{3} \right)$, and the corresponding graph of $W(n, L)$ versus L, with $k = 10^{-3.5}$m, $\alpha = 1/$m, is given in Figure 8.4. The open boxes represent the probability for $n = 30$, the solid circles for $n = 1500$, and the solid line for $n = 3000$. From the graph, we see that there are peaks for $W(n, L)$ corresponding to two different values of the total length L. As one extends the total length L, crossing the first maximum of $W(n, L)$ on the left shows that higher values of the winding number n are favored. One would then expect an overwinding event as L is increased in this domain. The situation is reversed as one continues to extend the length L and crosses the maximum on the right. Further extensions of L would favor the lower winding numbers n, and an unwinding of the macromolecule should be manifested [23].

8.8 Biomolecular Chirality and Entropy

A peculiar property of biomolecules is the apparent chiral asymmetry or preference. For example, amino acids in proteins of living organisms are left-handed, while all sugar molecules in the body are right-handed. There seems to be no *a priori* reason for such predominance of either left-handed or right-handed molecules. We can investigate analytically such chiral asymmetry by focusing on entropy differences associated with direction of winding of polymers. In particular, we consider helical polypeptides and determine the entropic difference between a left-handed helical protein and its right-handed mirror image.

Here, we calculate a general expression for the entropy S from the number of accessible conformational states for helical polymers. This is followed by a discussion on the explicit dependence of the probability distributions for clockwise and counterclockwise winding on the drift coefficient containing the interactions of the polymer with its immediate environment. We then show that an entropy difference ΔS arises between left-handed and right-handed winding distributions. For vanishing drift coefficient, the entropy difference also vanishes, recovering chiral symmetry.

We start by noting that helical polymers exhibit a circular topology with conformations winding about a given axis classified topologically. Homotopically inequivalent classes have assigned winding numbers [77; 185], $n = 0, \pm 1, \pm 2, ...$, such that a left-handed polymer can be designated as one which winds n-times counterclockwise $(n > 0)$ around the axis, and a right-handed polymer winds clockwise $|n|$-times with $n < 0$. The class $n = 0$, which signifies no winding, may be further distinguished. A polymer going clockwise without completing a turn is labelled by $n = 0^-$, and the one going counterclockwise with a turn less than 2π, by $n = 0^+$.

To get the entropy $S = k \ln \Omega$, we need to have the number of states Ω accessible to the system [96]. Consider then an ensemble of $\mathfrak{N}_0$ helical polymers. From this ensemble, let r_n be the number of polymers characterized by a particular winding number n, such that, $\sum_n r_n = \mathfrak{N}_0$. Suppose $W(n)$ is the probability that a polymer winds n-times. A single polymer with a particular number n of windings can be obtained from the ensemble with the probability, $\mathfrak{N}_0 W(n)$. A second polymer is obtained with a probability, $(\mathfrak{N}_0 - 1) W(n)$, the third with probability, $(\mathfrak{N}_0 - 2) W(n)$, and so on. From the remaining $(\mathfrak{N}_0 - r_n + 1)$ the r_nth polymer is obtained with probability, $(\mathfrak{N}_0 - r_n + 1) W(n)$. Hence, from a total of $\mathfrak{N}_0$ chains, with indistinguishable polymers in a given n-th class, the distinct number of ways of having r_n polymers characterized by a winding number n is,

$$\frac{\mathfrak{N}_0! \, (W(n))^{r_n}}{r_n! \, (\mathfrak{N}_0 - r_n)!} . \tag{8.28}$$

With this expression, we can obtain the number of ways of getting r_n polymers with winding number n, r_m polymers of winding number m, r_p

polymers of winding number p, and so on. This is given by,

$$\Omega = \left(\frac{\mathfrak{N}_0! \, (W(n))^{r_n}}{r_n! \, (\mathfrak{N}_0 - r_n)!} \right) \left(\frac{(\mathfrak{N}_0 - r_n)! \, (W(m))^{r_m}}{r_m! \, (\mathfrak{N}_0 - r_n - r_m)!} \right)$$

$$\times \left(\frac{(\mathfrak{N}_0 - r_n - r_m)! \, (W(p))^{r_p}}{r_p! \, (\mathfrak{N}_0 - r_n - r_m - r_p)!} \right) \cdots$$

which can be written as,

$$\Omega = \mathfrak{N}_0! \prod_{n=-\infty}^{+\infty} \frac{(W(n))^{r_n}}{(r_n)!} . \tag{8.29}$$

From Eq. (8.29), the entropy of the system, $S = k \ln \Omega$, can be evaluated to be,

$$S = k \ln \omega_r + k \sum_{n=-\infty}^{+\infty} r_n \ln W(n) , \tag{8.30}$$

where $\omega_r = \mathfrak{N}_0! \, (\prod_n (r_n)!)^{-1}$. In the next section, a specific form for the winding probability $W(n)$ is taken for polypeptides in solvents.

8.8.1 *Winding Probability*

A winding probability for polypeptides in solvents has been derived in an earlier section resulting in Eq. (8.19). For an arbitrary initial point, we let $\varphi_0 = \varphi_1$, and write,

$$W(n) = R\sqrt{\frac{4\pi}{Ll}} \frac{\exp\left[-\frac{R^2}{Ll} \left(2\pi n + \frac{l}{2DR} \int_0^L h(s) \, ds \right)^2 \right]}{\theta_3 \left(\frac{l}{4DR} \int_0^L h(s) \, ds \right)} , \tag{8.31}$$

where $\theta_3(u)$ is the theta function [98]. As has been observed, Eq. (8.31) shows that the winding probability depends explicitly on the drift coefficient $h(s)$, as s ranges from 0 to L along the length of a biopolymer.

An example of a length-dependent drift coefficient is of the form, $h(s) = \kappa \, J_\alpha(\nu s) \, J_{1-\alpha}(\nu(L-s))$, where $J_\alpha(\nu s)$ are Bessel functions, κ and ν are constants, and $0 \le s \le L$. Integrating $h(s)$ [3], and using this in Eq. (8.31)

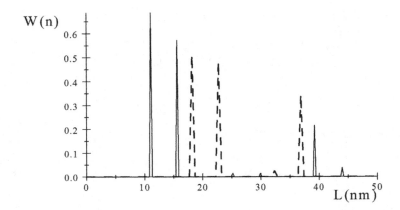

Fig. 8.5.   Solid line is for $n = 35$, Dashed line is for $n = -35$.

(with $\boldsymbol{\theta}_3(u) \approx 1$, for long polymers), the winding probability $W(n)$ varies with L differently for a right-handed polymer (clockwise winding) with $n = -35$, and a left-handed polymer (counterclockwise winding) with $n = 35$. For the values $R = 0.25$ nm, $l = 0.15$ nm which are typical of an α-helix, this variation is shown in Figure 8.5 where $\nu = 0.45/\text{nm}$, $(\kappa/D) = 740/\text{nm}$.

8.8.2   Chirality and Entropy

To evaluate the entropy for helical polymer windings and the corresponding mirror images, Eq. (8.29) is used,

$$\Omega^0 = \mathfrak{N}_0! \prod_{n=-\infty}^{+\infty} \frac{(W(n))^{r_n}}{(r_n)!} \, , \tag{8.32}$$

together with the conformations corresponding to the mirror-image described by,

$$\Omega^i = \mathfrak{N}_0! \prod_{n=-\infty}^{+\infty} \frac{(W(-n))^{r_n}}{(r_n)!} \, . \tag{8.33}$$

The entropy S^0 derived from Eq. (8.32) and entropy S^i from Eq. (8.33), allows the evaluation of the difference, $S^0 - S^i = \Delta S = k \ln(\Omega^0/\Omega^i)$. This yields,

$$\Delta S = k \sum_{n=-\infty}^{+\infty} r_n \ln\left(\frac{W(n)}{W(-n)}\right) . \tag{8.34}$$

Clearly, for systems where the equality $W(n) = W(-n)$ holds, one would have $\Delta S = 0$ in Eq. (8.34). However, substituting Eq. (8.31) into Eq. (8.34), a comparison of the values for left-handed ($n > 0$) and right-handed ($n < 0$) winding probabilities may be seen from the ratio,

$$\frac{W(n)}{W(-n)} = \exp\left[-\frac{4\pi nR}{LD}\int h(s)\, ds\right].\qquad (8.35)$$

The preferred handedness, or chirality, of a polymer is made manifest if the drift coefficient is nonzero, i.e., $h(s) \neq 0$ [28].

For helical polypeptides diffusing in a crowded biological cell one would expect a nonzero drift coefficient. Substitution of Eq. (8.35) in Eq. (8.34) yields the result,

$$\Delta S = -\left(\frac{k4\pi R}{LD}\int h(s)\; ds\right)\sum_{n=-\infty}^{+\infty} r_n n$$

$$\neq 0,\qquad (8.36)$$

where, in an ensemble, the number r_n of polymers with winding number n is not necessarily the same as the number r_{-n} of polymers with winding number $-n$, i.e., $|r_n n| \neq |(r_{-n})(-n)|$. This non-zero difference in entropy implies that a helical biopolymer has a different entropy from its mirror-image.

The results show a difference in entropy between mirror image distributions arising from the interaction of a biopolymer with its environment. This effect arises from the explicit dependence of the winding probability distributions for clockwise and counterclockwise winding on a drift coefficient varying with position along the polymer chain. This difference in entropy which depends on the chirality of winding could result in the predominance of polymers of one handedness over the other.

Chapter 9

White Noise Functional Integrals in Quantum Mechanics

We present an alternative method for evaluating the Feynman path integral based on white noise analysis. In particular, the Feynman integrand is identified as a generalized white noise functional, and an integration over the white noise measure yields the nonrelativistic quantum propagator. We start in the first two sections by introducing Feynman's formulation of non-relativistic quantum mechanics [83; 86], and its relation to the Schrödinger equation. The time-sliced form of the path integral approach is discussed as a procedure for arriving at exact expressions for the quantum mechanical propagator, from which one could extract the wave function and energy spectrum of the quantum system. We then look at the Hida-Streit white noise functional approach and end with some examples.

9.1 Quantum Propagator as Sum-Over-All-Histories

In quantum mechanics, the evolution of a state $|\Psi(t_0)\rangle$, at time t_0 to a state $|\Psi(t)\rangle$, where $t > t_0$ can be written as,

$$|\Psi(t)\rangle = U(t, t_0) \; |\Psi(t_0)\rangle, \qquad (9.1)$$

where $U(t, t_0)$ is the time-evolution operator. For a particle of mass μ described by the time-independent Hamiltonian operator, $H = (\mathbf{p}^2/2\mu) + V(\mathbf{r})$, the evolution operator has the form, $U(t, t_0) = \exp\{-(i/\hbar)H(t - t_0)\}$, and the time derivative of Eq. (9.1) leads to the Schrödinger equation,

$$H \; |\Psi(t)\rangle = i\hbar \frac{\partial}{\partial t} \; |\Psi(t)\rangle. \qquad (9.2)$$

In the coordinate representation, and using completeness of states,

$$1 = \int |\mathbf{r}'\rangle \langle \mathbf{r}'| \ d\mathbf{r}', \qquad (9.3)$$

we can write Eq. (9.2) as,

$$\int \langle \mathbf{r}| H |\mathbf{r}'\rangle \langle \mathbf{r}' | \Psi(t)\rangle \, d\mathbf{r}' = \frac{i}{\hbar} \frac{\partial}{\partial t} \langle \mathbf{r} | \Psi(t)\rangle, \qquad (9.4)$$

or,

$$\left[-\frac{\hbar^2}{2\mu} \boldsymbol{\nabla}^2 + V(\mathbf{r}) \right] \Psi(\mathbf{r}, t) = i\hbar \frac{\partial}{\partial t} \Psi(\mathbf{r}, t). \qquad (9.5)$$

We say that an exact solution to the differential equation (9.5) is possible if one obtains the wave function $\Psi(\mathbf{r}, t)$ in closed form as well as the energy spectrum E of the particle.

An integral approach to solving quantum mechanical problems is, however, also available. For a system with a time-independent Hamiltonian, let us again consider Eq. (9.1), i.e.,

$$|\Psi(t'')\rangle = \exp\left\{-(i/\hbar)H\tau\right\} \quad |\Psi(t')\rangle, \qquad (9.6)$$

where $\tau = (t'' - t')$. Writing this in the coordinate representation and with the help of Eq. (9.3), we have,

$$\Psi(\mathbf{r}'', t'') = \langle \mathbf{r}'' | \Psi(t'')\rangle = \langle \mathbf{r}''| \exp\left\{-(i/\hbar)H\tau\right\} |\Psi(t')\rangle$$

$$= \int \langle \mathbf{r}''| \exp\left\{-(i/\hbar)H\tau\right\} |\mathbf{r}'\rangle \langle \mathbf{r}' | \Psi(t')\rangle \ d\mathbf{r}'$$

$$\Psi(\mathbf{r}'', t'') = \int K(\mathbf{r}'', \mathbf{r}'; \tau) \Psi(\mathbf{r}', t') \ d\mathbf{r}', \qquad (9.7)$$

where the quantum propagator is,

$$K(\mathbf{r}'', \mathbf{r}'; \tau) = \langle \mathbf{r}''| \exp\left\{-(i/\hbar)H\tau\right\} |\mathbf{r}'\rangle. \qquad (9.8)$$

Equation (9.7) states that for a particle to evolve from a state $\Psi(\mathbf{r}', t')$ to a final state $\Psi(\mathbf{r}'', t'')$, one integrates over all possible values $\mathbf{r}'$ for $\Psi(\mathbf{r}', t')$ multiplied by the propagator $K(\mathbf{r}'', \mathbf{r}'; \tau)$.

The propagator (9.8) can also be written in another form which gives immediately the wave function Ψ and energy E of the system. We can write Eq. (9.8) as,

$$K\left(\mathbf{r''}, \mathbf{r'}; \tau\right) = \sum_{n,m} \langle \mathbf{r''}|m\rangle \langle m \mid \exp\left[-(i/\hbar)H\tau\right] \mid n\rangle \langle n|\mathbf{r'}\rangle$$

$$= \sum_{n} \Psi_n\left(\mathbf{r''}\right) \Psi_n^*\left(\mathbf{r'}\right) \exp\left[-(i/\hbar)E_n\tau\right], \qquad (9.9)$$

with

$$\Psi_m\left(\mathbf{r}\right) = \langle \mathbf{r}|m\rangle \qquad ; \qquad \langle m|n\rangle = \delta_{mn}\,. \qquad (9.10)$$

In 1948, Feynman proposed [83] a novel way of calculating the quantum mechanical propagator (9.8). Employing an idea from Dirac [71; 72], Feynman exploited the fact that for short intervals of time, the propagator has the form $K\left(t + \varepsilon; t\right) = (1/A) \exp\left[(i/\hbar) S\left(t + \varepsilon, t\right)\right]$, where $\varepsilon \ll 1$, and S is the classical action. As discussed in the next section, the equivalence of the Schrödinger equation (9.5) with the integral equation (9.7) can be used to show that one can indeed take this form for the short-time propagator. Hence, using the classical action, $S = \int L \, dt$, where L is the Lagrangian, Feynman allowed the particle to take all possible paths from $\mathbf{r'}$ to $\mathbf{r''}$, each path weighted by $\exp[(i/\hbar) S]$. All the possible histories, or trajectories, of the particle are summed up leading to Feynman's ansatz for the quantum mechanical propagator of the form,

$$K\left(\mathbf{r''}, \mathbf{r'}; \tau\right) = \sum_{paths} \exp[(i/\hbar) S]\,. \qquad (9.11)$$

A procedure that can be followed, therefore, starts by explicitly evaluating the path summation Eq. (9.11) and expressing it in the form of Eq. (9.9) where one can read off the wave function $\Psi_n(\mathbf{r})$ and the energy spectrum E_n. In this way, an exact solution of the quantum mechanical problem can be obtained without solving the Schrödinger equation (9.5) directly.

The approach of Feynman also highlights an essential difference between classical and quantum mechanics. In classical mechanics, an exact solution to a physical problem means that one is able to determine the exact path of the particle as it goes from a point $\mathbf{r'}$ at time t' to a point $\mathbf{r''}$ at a later time t''. Since classical mechanics is deterministic, there is only one classical path for the particle as embodied in Hamilton's Principle [95], $\delta S = 0$. In contrast, the path integral approach to quantum mechanics allows all possible paths, or histories, for the particle in going from $\mathbf{r'}$ to $\mathbf{r''}$, with each path weighted by, $\exp[(i/\hbar) S]$. A classical picture can then be obtained from quantum mechanics in the limit, $\hbar \to 0$, or $S \gg \hbar$, where contributions from non-classical paths add up to zero in the path integral [86].

9.2   Feynman Path Integral and the Schrödinger Equation

The Schrödinger equation (9.5) can be obtained [86] from the integral equation (9.7) by taking t'' to differ from t' by an infinitesimal amount ε. Let us set, $t' = t$, $t'' = t + \varepsilon$, $\mathbf{r}'' = \mathbf{r}$, and $\mathbf{r}' = \mathbf{r} - \boldsymbol{\eta}$ where $\boldsymbol{\eta} \ll 1$. Since integration is carried out only for $\mathbf{r}'$ in Eq. (9.7), we have $\int_{-\infty}^{\infty} d\mathbf{r}' = \int_{-\infty}^{\infty} d\boldsymbol{\eta}$, and this equation acquires the form,

$$\Psi(\mathbf{r}, t + \varepsilon) = \int K(\mathbf{r}, t + \varepsilon; \mathbf{r} - \boldsymbol{\eta}; t) \, \Psi(\mathbf{r} - \boldsymbol{\eta}, t) \, d\boldsymbol{\eta}. \tag{9.12}$$

For ε and $\boldsymbol{\eta}$ small, we have the following expansions,

$$\Psi(\mathbf{r}, t + \varepsilon) = \Psi(\mathbf{r}, t) + \varepsilon \frac{\partial}{\partial t} \Psi(\mathbf{r}, t) + \frac{1}{2} \varepsilon^2 \frac{\partial^2}{\partial t^2} \Psi(\mathbf{r}, t) + \dots \tag{9.13}$$

$$\Psi(\mathbf{r} - \boldsymbol{\eta}, t) = \Psi(\mathbf{r}, t) - \sum_{\alpha=1}^{3} \eta_\alpha \nabla_\alpha \Psi(\mathbf{r}, t)$$

$$+ \frac{1}{2} \sum_{\alpha, \beta=1}^{3} \eta_\alpha \eta_\beta \nabla_\alpha \nabla_\beta \Psi(\mathbf{r}, t) + \dots. \tag{9.14}$$

Utilizing an idea of Dirac [71; 72], Feynman expressed the propagator for an infinitesimal time-interval as,

$$K(\mathbf{r}, t + \varepsilon; \mathbf{r} - \boldsymbol{\eta}; t) = (1/A) \exp\left[(i/\hbar) S(t + \varepsilon, t) \right], \tag{9.15}$$

where $S(t + \varepsilon, t) \approx \varepsilon \, L(\boldsymbol{\eta}/\varepsilon, \mathbf{r})$, is the short-time action and L the Lagrangian. This equation can be used in Eq. (9.15) by writing it as,

$$K(\mathbf{r}, t + \varepsilon; \mathbf{r} - \boldsymbol{\eta}; t) = (1/A) \exp\left[(i\mu\boldsymbol{\eta}^2/2\hbar\varepsilon) - i\varepsilon V(\mathbf{r})/\hbar \right]$$

$$= (1/A) \exp\left[(i\mu\boldsymbol{\eta}^2/2\hbar\varepsilon) \right]$$

$$\times \left\{ 1 - i\varepsilon V(\mathbf{r})/\hbar + \dots \right\}. \tag{9.16}$$

With Eqs. (9.13)–(9.16), Eq. (9.12) becomes,

$$\Psi(\mathbf{r},t) + \varepsilon\frac{\partial}{\partial t}\Psi(\mathbf{r},t) + \frac{1}{2}\varepsilon^2\frac{\partial^2}{\partial t^2}\Psi(\mathbf{r},t) + \dots$$

$$= \int d\boldsymbol{\eta}\ (1/A)\exp\left[(i\mu\boldsymbol{\eta}^2/2\hbar\varepsilon)\right]\{1 - i\varepsilon V(\mathbf{r})/\hbar + \dots\}$$

$$\times\left\{\Psi(\mathbf{r},t) - \sum_{\alpha=1}^{3}\eta_\alpha\nabla_\alpha\Psi(\mathbf{r},t) + \frac{1}{2}\sum_{\alpha,\beta=1}^{3}\eta_\alpha\eta_\beta\nabla_\alpha\nabla_\beta\Psi(\mathbf{r},t) + \dots\right\}.$$

(9.17)

The integration can be performed with the following results [83; 86],

$$\int\limits_{-\infty}^{\infty}\int\int\exp\left(i\mu\boldsymbol{\eta}^2/2\hbar\varepsilon\right)d\boldsymbol{\eta} = (2\pi i\hbar\varepsilon/\mu)^{3/2}\,;$$

(9.18)

$$\int\limits_{-\infty}^{\infty}\int\int\exp\left(i\mu\boldsymbol{\eta}^2/2\hbar\varepsilon\right)\eta_\alpha\,d\boldsymbol{\eta} = 0$$

(9.19)

$$\int\limits_{-\infty}^{\infty}\int\int\exp\left(i\mu\boldsymbol{\eta}^2/2\hbar\varepsilon\right)\eta_\alpha\eta_\beta\,d\boldsymbol{\eta} = (i\hbar\varepsilon/\mu)\,(2\pi i\hbar\varepsilon/\mu)^{3/2}\,\delta_{\alpha\beta}\,.$$

(9.20)

Note that Eqs. (9.18) to (9.20) are oscillatory integrals which may be defined by putting in a small imaginary number $i\delta$ where $0 < \delta \ll 1$ (e.g., by replacing $\hbar$ with $\hbar(1 - i\delta)$), and then taking the limit $\delta \to 0$. Equations (9.18) to (9.20) allow us to write Eq. (9.17) as,

$$\Psi(\mathbf{r},t) + \varepsilon\frac{\partial}{\partial t}\Psi(\mathbf{r},t) + \dots = (1/A)\,(2\pi i\hbar\varepsilon/\mu)^{3/2}$$

$$\times\{\Psi(\mathbf{r},t) - i\varepsilon V(\mathbf{r})\,\Psi(\mathbf{r},t)/\hbar\}$$

$$+ (i\hbar\varepsilon/2A\mu)\,(2\pi i\hbar\varepsilon/\mu)^{3/2}\sum_{\alpha,\beta=1}^{3}\delta_{\alpha\beta}\nabla_\alpha\nabla_\beta\Psi(\mathbf{r},t) + \dots.$$

(9.21)

Choosing the normalization to be, $A = (2\pi i\hbar\varepsilon/\mu)^{3/2}$, and equating terms of order ε, Eq. (9.21) reduces to the Schrödinger equation, Eq. (9.5). We note that a crucial link in showing that the integral equation (9.7) is equivalent to the Schrödinger equation is the expression for the short-time propagator Eq. (9.15).

9.3   The Path Integral in Time-Sliced Form

In the actual evaluation of the propagator as the path integral Eq. (9.11), it is often convenient to use its time sliced form. This can be obtained from the definition of the quantum propagator Eq. (9.8) by slicing the time τ into N-subintervals, i.e., $\tau/N = \varepsilon_j$, such that,

$$\tau = \sum_{j=1}^{N} \varepsilon_j \,. \tag{9.22}$$

Equation (9.8) can then be written as,

$$\begin{aligned}
K\left(\mathbf{r}'', \mathbf{r}'; \tau\right) &= \left\langle \mathbf{r}'' | e^{-(i/\hbar)H\varepsilon_N} \cdot e^{-(i/\hbar)H\varepsilon_{N-1}} ... e^{-(i/\hbar)H\varepsilon_1} | \mathbf{r}' \right\rangle \\
&= \int ... \int \left\langle \mathbf{r}_N | e^{-(i/\hbar)H\varepsilon_N} | \mathbf{r}_{N-1} \right\rangle \\
&\quad \times \left\langle \mathbf{r}_{N-1} | e^{-(i/\hbar)H\varepsilon_{N-1}} | \mathbf{r}_{N-2} \right\rangle ... \\
&\quad \times \left\langle \mathbf{r}_1 | e^{-(i/\hbar)H\varepsilon_1} | \mathbf{r}_0 \right\rangle d\mathbf{r}_1 ... d\mathbf{r}_{N-1} \,,
\end{aligned} \tag{9.23}$$

where completeness of states Eq. (9.3) has been utilized, and $\mathbf{r}_j = \mathbf{r}(\varepsilon_j)$, with the endpoints being $\mathbf{r}'' = \mathbf{r}_N$ and $\mathbf{r}' = \mathbf{r}_0$. Now, for each time interval ε_j we employ the relation Eq. (9.15) such that,

$$\left\langle \mathbf{r}_j | e^{-(i/\hbar)H\varepsilon_j} | \mathbf{r}_{j-1} \right\rangle = (1/A) \, e^{(i/\hbar)S_j} \,, \tag{9.24}$$

where S_j is the short-time action, which allows us to write Eq. (9.23) as,

$$\begin{aligned}
K\left(\mathbf{r}'', \mathbf{r}'; \tau\right) &= \int ... \int \frac{e^{(i/\hbar)S_N}}{A} \frac{e^{(i/\hbar)S_{N-1}}}{A} \quad ... \quad \frac{e^{(i/\hbar)S_1}}{A} \\
&\quad \times \quad d\mathbf{r}_1 ... d\mathbf{r}_{N-1} \,.
\end{aligned} \tag{9.25}$$

The time-sliced form of the propagator is obtained by letting, $\varepsilon_j \to 0$, $N \to \infty$ and $\tau = N\varepsilon_j$, i.e.,

$$K\left(\mathbf{r}'', \mathbf{r}'; \tau\right) = \lim_{N \to \infty} \int \prod_{j=1}^{N} \exp\left[(i/\hbar)\, S_j\right] \prod_{j=1}^{N} \left(\mu/2\pi i\hbar\varepsilon_j\right)^{3/2} \prod_{j=1}^{N-1} \left(d\mathbf{r}_j\right) \,. \tag{9.26}$$

The normalization A is chosen as, $A = \left(2\pi i\hbar\varepsilon/\mu\right)^{3/2}$, so that the propagator satisfies the initial boundary condition

$$\lim_{\tau \to 0} K\left(\mathbf{r}'', \mathbf{r}'; \tau\right) = \delta\left(\mathbf{r}'' - \mathbf{r}'\right) \,. \tag{9.27}$$

Symbolically, Eq. (9.26) can be written as Eq. (9.11). Graphically, it can be represented by a summation, or integration, over all trajectories of the particle from $\mathbf{r}'$ to $\mathbf{r}''$ as shown in Figure 3.1.

The short-time action in Eq. (9.26) is approximated as,

$$S_j = \int_{t_{j-1}}^{t_j} \left[\frac{1}{2}\mu \, \dot{\mathbf{r}}^2 - V(\mathbf{r}) \right] dt = \frac{\mu (\triangle \mathbf{r}_j)^2}{2\varepsilon_j} - V(\mathbf{r}_j)\varepsilon_j, \qquad (9.28)$$

where $(\triangle \mathbf{r}_j)^2 = (\mathbf{r}_j - \mathbf{r}_{j-1})^2$ and $\varepsilon_j = t_j - t_{j-1}$. Once the form of the potential $V(\mathbf{r})$ is specified, the path integral Eq. (9.26) may be evaluated. It is important to remember, however, that terms in the short-time action of order $O\left(\varepsilon_j^{1+\delta}\right)$, where $\delta > 0$, have negligible contributions and, hence, can be ignored. The relation, $(\triangle x_j)^2 \approx \varepsilon_j$, often comes in handy in determining the terms to be dropped [152; 185] where x is the coordinate.

We note at this point that although the application and usefulness of the path integral in physics are beyond doubt [129], a rigorous mathematical proof for the existence of the Feynman integral Eq. (9.26) remains to be fully explored [7; 68; 125; 196]. Numerous quantum mechanical problems are solved using the time slicing procedure [86; 99], but the path integral has often been criticized for its lack of mathematical meaning. Specifically the integral, Eq. (9.26), with its infinite-dimensional flat "measure" $\lim_{N \to \infty} \prod (d\mathbf{r}_j)$ is not mathematically well-defined. It has been observed, however, that if one performs an analytic continuation replacing the time t by $-it$, the measure in the Feynman integral becomes the well-defined Wiener measure. In this case, the Schrödinger equation is transformed into the heat equation with $K(\mathbf{r}'', \mathbf{r}'; \tau)$ as a solution. There have been several attempts to provide the Feynman integral with a more solid mathematical foundation [7; 68; 69; 70; 125; 126; 127; 196], but we discuss here an approach pioneered in 1983 by Streit and Hida [139; 196] which utilizes the infinite dimensional white noise calculus [106; 133; 163].

9.4   The Free Particle

Let us take a free particle of mass μ as our first example in explicitly evaluating the propagator $K(\mathbf{r}'', \mathbf{r}'; \tau)$ using the time slicing procedure. With $V(\mathbf{r}) = 0$, the short-time action Eq. (9.28) becomes,

$$S_j = \frac{\mu}{2\varepsilon_j} \left[(\triangle x_j)^2 + (\triangle y_j)^2 + (\triangle z_j)^2 \right], \qquad (9.29)$$

where $\mathbf{r} = (x, y, z)$. The exponential of this action, i.e.,

$$\exp\left[(i/\hbar)\, S_j\right] = \exp\left[(i/\hbar)\, S_j(x)\right]\ \exp\left[(i/\hbar)\, S_j(y)\right]\ \exp\left[(i/\hbar)\, S_j(z)\right], \tag{9.30}$$

allows the propagator Eq. (9.26) to be separated into its x, y and z components such that we can write,

$$K\left(\mathbf{r}'', \mathbf{r}'; \tau\right) = K\left(x'', x'; \tau\right)\ K(y'', y'; \tau)\ K\left(z'', z'; \tau\right). \tag{9.31}$$

From Eq. (9.26), the $K\left(x'', x'; \tau\right)$ is given by,

$$K\left(x'', x'; \tau\right) = \lim_{N \to \infty} \int \prod_{j=1}^{N} \exp\left[(i/\hbar)\, S_j(x)\right] \prod_{j=1}^{N} \left(\mu/2\pi i\hbar\varepsilon_j\right)^{1/2} \prod_{j=1}^{N-1} \left(dx_j\right), \tag{9.32}$$

where $S_j(x) = \mu(\triangle\, x_j)^2/2\varepsilon_j$, and $(\triangle x_j)^2 = (x_j - x_{j-1})^2$. The $K(y'', y'; \tau)$ and $K\left(z'', z'; \tau\right)$ have forms identical to Eq. (9.32), and it is sufficient to show only the path integration of the x-component of the full propagator. Explicitly, Eq. (9.32) appears as,

$$\begin{aligned} K\left(x'', x'; \tau\right) = \lim_{N \to \infty} \left(\mu/2\pi i\hbar\varepsilon\right)^{N/2} \int \dots \int \exp\left[i\mu\left(x_1 - x_0\right)^2/2\hbar\varepsilon\right] \\ \times \exp\left[i\mu\left(x_2 - x_1\right)^2/2\hbar\varepsilon\right] \dots \\ \times \exp\left[i\mu\left(x_N - x_{N-1}\right)^2/2\hbar\varepsilon\right] dx_1 dx_2 \cdots dx_{N-1}, \end{aligned} \tag{9.33}$$

where $\varepsilon_j = \varepsilon$ $(j = 1, ..., N)$. The series of integration to be performed is facilitated by the identity [86; 185],

$$\begin{aligned} \int\limits_{-\infty}^{+\infty} du\sqrt{a/\pi} \exp\left[-a\left(x - u\right)^2\right] \sqrt{b/\pi} \exp\left[-b\left(u - y\right)^2\right] \\ = \sqrt{ab/\pi\left(a + b\right)} \exp\left[-ab\left(x - y\right)^2/\left(a + b\right)\right]. \end{aligned} \tag{9.34}$$

With Eq. (9.34) the integral over x_1 for the first time interval yields,

$$\begin{aligned} \int dx_1 \sqrt{\mu/2\pi i\hbar\varepsilon} \exp\left[i\mu\left(x_1 - x_0\right)^2/2\hbar\varepsilon\right] \\ \times \sqrt{\mu/2\pi i\hbar\varepsilon} \exp\left[i\mu\left(x_2 - x_1\right)^2/2\hbar\varepsilon\right] \\ = \sqrt{\mu/2\pi i\hbar\left(2\varepsilon\right)} \exp\left[i\mu\left(x_2 - x_0\right)^2/2\hbar\left(2\varepsilon\right)\right]. \end{aligned} \tag{9.35}$$

Likewise, the integral over x_2 yields,

$$\int dx_2 \sqrt{\mu/2\pi i\hbar\,(2\varepsilon)} \exp\left[i\mu\,(x_2 - x_0)^2 / 2\hbar\,(2\varepsilon)\right]$$

$$\times \sqrt{\mu/2\pi i\hbar\varepsilon} \exp\left[i\mu\,(x_3 - x_2)^2 / 2\hbar\varepsilon\right]$$

$$= \sqrt{\mu/2\pi i\hbar\,(3\varepsilon)} \exp\left[i\mu\,(x_3 - x_0)^2 / 2\hbar\,(3\varepsilon)\right], \qquad (9.36)$$

and an obvious pattern appears as a result of each integration. The last integral over x_{N-1}, therefore, gives us,

$$K\,(x'', x'; \tau) = \lim_{N \to \infty} \sqrt{\mu/2\pi i\hbar\,(N\varepsilon)} \exp\left[i\mu\,(x_N - x_0)^2 / 2\hbar\,(N\varepsilon)\right]. \quad (9.37)$$

With $N\varepsilon = \tau$, the final form of the propagator for motion along the x-coordinate is,

$$K\,(x'', x'; \tau) = \sqrt{\mu/2\pi i\hbar\tau} \exp\left[i\mu\,(x'' - x')^2 / 2\hbar\tau\right], \qquad (9.38)$$

where $x'' = x_N$ and $x' = x_0$. Since $K\,(y'', y'; \tau)$ and $K\,(z'', z'; \tau)$ acquire the same form as Eq. (9.38), the full propagator Eq. (9.31) is simply,

$$K\,(\mathbf{r''}, \mathbf{r'}; \tau) = (\mu/2\pi i\hbar\tau)^{3/2} \exp\left[i\mu\,(\mathbf{r''} - \mathbf{r'})^2 / 2\hbar\tau\right]. \qquad (9.39)$$

To extract the wave functions and energy spectrum, we use the Gaussian integral $(a > 0)$,

$$\int\limits_{-\infty}^{+\infty} \exp\left(-\frac{i}{2}ap^2 - ibp\right) dp = \sqrt{2\pi/ia} \exp\left(ib^2/2a\right), \qquad (9.40)$$

to rewrite Eq. (9.39) as,

$$K\,(\mathbf{r''}, \mathbf{r'}; \tau) = \int \Psi\,(\mathbf{r''})^* \, \Psi\,(\mathbf{r'}) \exp\left[-iE\tau/\hbar\right] d^3\mathbf{p}. \qquad (9.41)$$

This gives us the energy, $E = \hbar^2 p^2 / 2\mu$, and the wave function,

$$\Psi\,(\mathbf{r}) = (2\pi)^{-3/2} \exp\left(i\mathbf{p} \cdot \mathbf{r}\right). \qquad (9.42)$$

This free particle example in Cartesian coordinates is the simplest case with which one can demonstrate exact path integration to obtain the wave function and energy spectrum. In the next section we discuss an alternative way of evaluating the propagator using white noise analysis.

9.5   Feynman Integrand as a White Noise Functional

Here we give a brief review of how the Feynman integral in quantum mechanics can be cast as an integration of a white noise functional over the Gaussian white noise measure $d\mu$. Evaluation of the appropriate white noise functional using $d\mu$ then readily yields the quantum propagator [196]. Since Feynman introduced his path integral formulation of quantum mechanics [83], various approaches aimed at providing a mathematically rigorous meaning to the path integral have been given. Notable among these would be the prodistributions of DeWitt-Morette [70], the oscillatory integrals of Albeverio and Høegh-Krohn [7] which explore the Fresnel integrals, and the white noise analysis approach of Hida and Streit [196]. We shall consider the Hida-Streit approach which, through the years, has allowed the solution of various classes of potentials. Note that this approach differs from the use of white noise in Parisi-Wu stochastic quantization in field theory [168].

Aside from being mathematically well-defined, the merits of the white noise approach include the following:

(a) The imaginary $i = \sqrt{-1}$ is retained where Feynman originally placed it, i.e., $\exp(iS)$. We note that the usual procedure of performing an analytic continuation to imaginary time to supply mathematical rigor to the Feynman integral leads to a loss (together with the i) of some important quantum features such as quantum interference.

(b) The time-slicing procedure, Eq. (9.26), becomes unnecessary. One of the drawbacks of the time-slicing technique in evaluating the Feynman integral is the ambiguity introduced in choosing whether the value of the variables should be taken at the pre-point t_{j-1}, the mid-point $(t_j + t_{j-1})/2$, or the geometric mean $(t_j t_{j-1})^{1/2}$, of the small time interval.

(c) Varied quantum mechanical problems could actually be solved using this procedure. Among the solvable quantum systems are the harmonic [196] and anharmonic oscillator [100], particle in a uniform magnetic field [61], the Morse potential [132], a particle in a circle [21; 140], the Aharonov-Bohm set-up [21], particle in a box [21], and time-dependent potentials [100].

We now proceed to illustrate the Hida-Streit formalism and cast the Feynman path integral in the language of white noise analysis [196; 197]. The summation-over-all histories approach introduced by R. Feynman expresses the quantum propagator as an integral over all possible paths symbolically written as,

$$K\left(x_1, x_0; \tau\right) = \int \exp\left(\frac{i}{\hbar}S\right) \mathcal{D}\left[x\right], \tag{9.43}$$

which is a continuum version of Eq. (9.11). The path $x(\tau)$ of a particle, which starts from an initial point x_0 is then parametrized in terms of the Brownian motion $B(\tau)$ as,

$$x(\tau) = x_0 + \sqrt{\hbar/\mu}\ B(\tau)$$
$$= x_0 + \sqrt{\hbar/\mu} \int_0^\tau \omega(t)\, dt\,. \tag{9.44}$$

This means we let, $f\left(\tau - t\right) = h\left(t\right) = 1$, in Eq. (3.6). With this, the velocity of the particle becomes, $(dx/dt) = \sqrt{\hbar/\mu}\ \omega$, which enables us to write the exponential of $(i/\hbar)S_0$, where S_0 is the action for the free particle, as

$$\exp\left(\frac{i}{\hbar}S_0\right) = \exp\left[\frac{i}{\hbar}\int_0^\tau \left(\frac{1}{2}\mu \dot{x}^2\right) dt\right]$$
$$= \exp\left[\frac{i}{2}\int_0^\tau \omega(t)^2\, dt\right]. \tag{9.45}$$

Having expressed $\exp[(i/\hbar)\, S_0]$ in terms of the random white noise variable $\omega(t)$, the integration over all paths ($\lim_{N\to\infty} \prod(dx_j)$ or $d^\infty x$) leads to an integration over the Gaussian white noise measure $d\mu(\omega)$. In particular, we have the correspondence, $\lim_{N\to\infty} \prod^N (A_j) \prod^{N-1} (dx_j) \to N_\omega\, d^\infty\omega$. From Eq. (2.7) we have, $N_\omega\ d^\infty\omega = \exp\left(\frac{1}{2}\int \omega(t)^2\, dt\right) d\mu(\omega)$, and the exponential factor $\exp\left(\frac{1}{2}\int \omega(t)^2\, dt\right)$ combines with Eq. (9.45) giving the Gauss kernel,

$$I_0 = \mathcal{N} \exp\left[\left(\frac{i+1}{2}\right)\int_0^\tau \omega(t)^2\, dt\right], \tag{9.46}$$

where $\mathcal{N}$ is the normalization.

We then note that the paths in the Feynman integral begin at x_0 and end at x_1. However, the parametrization of the paths $x(\tau)$ in Eq. (9.44) shows that only the initial point x_0 is fixed from where the random Brownian motion begins. We, therefore, fix the endpoint with a Donsker delta function, $\delta\left(x(\tau)-x_1\right)$, such that at time τ the particle is at x_1. The Feynman integrand can then be represented by,

$$I = I_0\,\delta\left(x(\tau)-x_1\right), \qquad (9.47)$$

where $x(\tau)$ and I_0 are given by Eqs. (9.44) and (9.46), respectively. Mathematically, Eq. (9.47) has been shown to exist as a Hida distribution [62]. We thus arrive at the correspondence, for the free particle, of the Feynman integration over paths and the infinite-dimensional white noise integral,

$$\int \exp\left(iS_0\right)\,D[x] \equiv \mathcal{N}\int \exp\left[\left(\frac{i+1}{2}\right)\int_0^\tau \omega(t)^2\,dt\right]\delta\left(x(\tau)-x_1\right)\,d\mu(\omega).$$
$$(9.48)$$

The presence of an external potential $V(x)$ can also be accommodated into the formalism [122]. As can be seen from Eqs. (9.26) and (9.28), an external potential modifies the Feynman integrand, Eq. (9.47), such that the corresponding white noise functional to be considered becomes,

$$I_V = I_0\,\delta\left(x(\tau)-x_1\right)\exp\left[-\frac{i}{\hbar}\int_0^\tau V\left(x(t)\right)\,dt\right]. \qquad (9.49)$$

The product $\Phi\cdot\delta(\langle\omega,g\rangle-a)$, of a Hida distribution $\Phi(\omega)$ with a Donsker delta function may be handled by first expressing the delta function as in Eq. (2.29), and getting the T-transform of the product using Eq. (2.8) to get,

$$T\left(\Phi\cdot\delta\left(\langle\omega,g\rangle-a\right)\right)(\xi) = T\left(\frac{1}{2\pi}\int\Phi\exp\left[ip\left(\langle\omega,g\rangle-a\right)\right]dp\right)(\xi)$$
$$= \frac{1}{2\pi}\int\int\Phi\exp\left(i\langle\omega,\xi\rangle\right)\exp\left[ip\left(\langle\omega,g\rangle-a\right)\right]$$
$$\times dp\,d\mu(\omega). \qquad (9.50)$$

Since the T-transform and integration over p commute, we get a useful formula [60; 100],

$$T\left(\Phi\cdot\delta\left(\langle\omega,g\rangle-a\right)\right)(\xi) = \frac{1}{2\pi}\int\exp(-ipa)T\Phi(\xi+pg)\,dp. \qquad (9.51)$$

To facilitate practical applications, we now proceed to obtain the propagator of some quantum problems.

9.5.1   *Free Particle*

Having expressed the free particle Feynman integrand as a white noise functional given by Eq. (9.47), we now write the delta function as in Eq. (2.29) to express I as,

$$I = \frac{1}{2\pi} \int \exp\left[ip(x_0 - x_1)\right] \exp\left[ip\sqrt{\hbar/\mu} \int \omega \, dt\right] I_0 \, dp. \tag{9.52}$$

Integrating over $d\mu(\omega)$ we have,

$$
\begin{aligned}
K(x_1, x_0; \tau) &= \int I \, d\mu(\omega) \\
&= \frac{1}{2\pi} \int dp \exp\left[ip(x_0 - x_1)\right] \\
&\quad \times \int \exp\left[\left(ip\sqrt{\hbar/\mu}\right) \int \omega \, dt\right] I_0 \, d\mu(\omega) \\
&= \frac{1}{2\pi} \int dp \exp\left[ip(x_0 - x_1)\right] T I_0\left(p\sqrt{\hbar/\mu}\right), \tag{9.53}
\end{aligned}
$$

where we identified the integral over $d\mu(\omega)$ as the T-transform, $T I_0(\xi)$, with $\xi = p\sqrt{\hbar/\mu}$ (see Eq. (2.8)). Using Eq. (2.28) for the T-transform of I_0 we get,

$$
\begin{aligned}
K(x_1, x_0; \tau) &= \frac{1}{2\pi} \int dp \exp\left[ip(x_0 - x_1)\right] \exp\left[-\frac{i}{2} \int_0^\tau \left(p\sqrt{\hbar/\mu}\right)^2 dt\right] \\
&= \frac{1}{2\pi} \int dp \exp\left[ip(x_0 - x_1)\right] \exp\left[-i\hbar p^2 \tau 2\mu\right] \\
&= \int \Psi(x_0) \, \Psi(x_1)^* \exp(-iE\tau/\hbar) \, dp, \tag{9.54}
\end{aligned}
$$

where $\Psi(x) = (1/\sqrt{2\pi}) \exp(ipx)$, and $E = \hbar^2 p^2/2\mu$, are the wave function and energy of the free particle, respectively.

9.5.2   *Velocity-dependent Potential: $V_1 = -\dot{x}\xi(t)$*

Using the parametrization Eq. (9.44), we have, $\dot{x} = \sqrt{\hbar/\mu}\, \omega$ and $V_1 = -\sqrt{\hbar/\mu}\, \omega(t)\xi(t)$. With this potential, Eq. (9.49) becomes,

$$I_{V_1} = I_0 \, \delta\left(x(\tau) - x_1\right) \exp\left[\frac{i}{\sqrt{\mu\hbar}} \int_0^\tau \omega(t)\xi(t) \, dt\right]. \tag{9.55}$$

To get the propagator, we integrate Eq. (9.55) over $d\mu(\omega)$, and note that this is similar to the T-transform of I, Eq. (9.47), i.e.,

$$
K_{V_1}(x_1, x_0; \tau) = \int I_{V_1} d\mu(\omega) = TI\left(\xi/\sqrt{\mu\hbar}\right)
$$

$$
= T\left(I_0\,\delta\left(\sqrt{\hbar/\mu}\int \omega\,dt - (x_1 - x_0)\right)\right)\left(\xi/\sqrt{\mu\hbar}\right)
$$

$$
= \frac{1}{2\pi}\int \exp\left[-ip(x_1 - x_0)\right] TI_0\left(\frac{\xi}{\sqrt{\mu\hbar}} + \sqrt{\frac{\hbar}{\mu}}p\right) dp,
$$
(9.56)

where we employed the formula Eq. (9.51). Using Eq. (2.28) to evaluate TI_0, we get,

$$
K_{V_1}(x_1, x_0; \tau) = \frac{1}{2\pi}\int \exp\left[-ip(x_1 - x_0)\right]
$$

$$
\times \exp\left(-\frac{i}{2}\int_0^\tau \left(\frac{\xi}{\sqrt{\mu\hbar}} + \sqrt{\frac{\hbar}{\mu}}p\right)^2 dt\right) dp
$$

$$
= \frac{1}{2\pi}\exp\left(-\frac{i}{2\mu\hbar}\int_0^\tau \xi^2 dt\right)
$$

$$
\times \int \exp\left\{-\frac{i\hbar\tau p^2}{2\mu} - i\left[\frac{1}{\mu}\int_0^\tau \xi\,dt + (x_1 - x_0)\right]p\right\} dp.
$$
(9.57)

Integrating the Gaussian integral over p, we obtain the propagator for the velocity-dependent potential,

$$
K_{V_1}(x_1, x_0; \tau) = \sqrt{\frac{\mu}{2\pi i\hbar\tau}}\exp\left\{\left(-\frac{i}{2\mu\hbar}\int_0^\tau \xi^2 dt\right)\right.
$$

$$
\left. + \frac{i\mu}{2\hbar\tau}\left[(1/\mu)\int_0^\tau \xi(t)dt + (x_1 - x_0)\right]^2\right\}.
$$
(9.58)

Note that when $\xi = 0$, we get the free particle propagator.

9.5.3 *Time-dependent Potential: $V_2 = \dot{\xi}x$*

The propagator for this case is readily derived from Eq. (9.58) for the potential, $V_1 = -\sqrt{\hbar/\mu}\,\omega(\tau)\xi(\tau)$. In fact, one can relate $\omega(t)\xi(t)$ with the potential, $V_2 = \dot{\xi}x$. Since $\omega(t)$ is given by Eq. (2.3) we can write,

$$i \int_0^\tau \omega(t)\xi(t)\, dt = i \int_0^\tau \frac{d}{dt}\,[B(t)\xi(t)]\, dt - i \int_0^\tau B(t)\frac{d}{dt}\xi(t)\, dt$$

$$= iB(\tau)\xi(\tau) - iB(0)\xi(0) - i \int_0^\tau B(t)\dot{\xi}dt. \qquad (9.59)$$

From Eq. (9.44), we have, $B(\tau) = \sqrt{\mu/\hbar}[x(\tau) - x_0]$, and Eq. (9.59) becomes,

$$i \int_0^\tau \omega(t)\xi(t)\, dt = -i\sqrt{\mu/\hbar} \int_0^\tau \dot{\xi}x(t)\, dt + i\sqrt{\mu/\hbar}\,[\xi(\tau)\,x(\tau) - \xi(0)\,x_0].$$

$$(9.60)$$

Using this together with Eqs. (9.55) and (9.56) of the previous example, we see that,

$$K_{V_1}(x_1, x_0; \tau) = \int I_0\, \delta\,(x(\tau) - x_1)\, \exp\left(-\frac{i}{\hbar} \int_0^\tau \dot{\xi}x(t)\, dt\right)\, d\mu(\omega)$$

$$\times\, \exp\left(\frac{i}{\hbar}\,[\xi(\tau)\,x(\tau) - \xi(0)\,x_0]\right). \qquad (9.61)$$

The second exponential factor is in terms of the initial and final values of ξ and x and can be transferred to the other side of the equation. Hence, using Eq. (9.61), we get an expression for the propagator with potential $V_2 = \dot{\xi}x$, i.e.,

$$K_{V_2}(x_1, x_0; \tau) = \int I_0\, \delta\,(x(\tau) - x_2)\, \exp\left(-\frac{i}{\hbar} \int_0^\tau \dot{\xi}x(t)\, dt\right)\, d\mu(\omega)$$

$$= K_{V_1}(x_1, x_0; \tau) \exp\left(\frac{i}{\hbar}\,[\xi(0)\,x_0 - \xi(\tau)\,x(\tau)]\right), \qquad (9.62)$$

where $K_{V_1}(x_1, x_0; \tau)$ is given by Eq. (9.58). For example, from Eq. (9.62), the propagator for the potential $V = kx$ can readily be obtained by choosing $\xi = kt$.

9.6 Uniform Magnetic Field

Uniform magnetic fields arise in many interesting physical systems. As an application of the white noise treatment of Lévy's stochastic area (see

Section 2.5), we now consider a particle subjected to a uniform magnetic field oriented along the z-axis [61]. We represent the magnetic field by, $\mathbb{B} = \mathcal{B}\hat{k}$, to distinguish it from our notation for the Brownian motion, $B(t)$. The vector potential for this case is given by $\mathbf{A} = \left(-\frac{\mathcal{B}}{2}y, \frac{\mathcal{B}}{2}x, 0\right)$. Since motion along the z-axis is that of a free particle, we shall focus our discussion on the $(x - y)$-plane. The classical action can be written as (we set $\mu = \hbar = c = 1$),

$$
S = \int\limits_0^T \left[\frac{1}{2}\dot{\mathbf{r}}^2 + e\mathbf{A}\cdot\dot{\mathbf{r}}\right] dt
$$

$$
= \int\limits_0^T \left[\frac{1}{2}\left(\dot{x}^2 + \dot{y}^2\right)\right] dt
$$

$$
+ (e\mathcal{B}/2) \int\limits_0^T \left(x\dot{y} - y\dot{x}\right) dt\,, \qquad (9.63)
$$

where $\mathbf{r} = (x, y)$. Using this in $\exp{(iS)}$, the propagator becomes,

$$
K\left(x_1, y_1; x_0, y_0\right) = \int \exp\left(\frac{i}{2}\int\limits_0^T \left(\dot{x}^2 + \dot{y}^2\right) dt\right)
$$

$$
\times \exp\left[i\gamma \int\limits_0^T \left(x\dot{y} - y\dot{x}\right) dt\right] D\left[xy\right]\,, \qquad (9.64)
$$

where $\gamma = e\mathcal{B}/2$. We now evaluate $K\left(x_1, y_1; x_0, y_0\right)$ by parametrizing the paths in terms of Brownian motion variables as,

$$
x(t) = x_0 + B_x(t)
$$

$$
= x_0 + \int\limits_0^t \omega_x(\tau)\, d\tau\,, \qquad (9.65)
$$

where $\omega_x(t) = dB_x(t)/dt$, is the corresponding white noise variable. Similarly,

$$
y(t) = y_0 + B_y(t)
$$

$$
= y_0 + \int\limits_0^t \omega_y(\tau)\, d\tau\,, \qquad (9.66)
$$

with $\omega_y(t) = dB_y(t)/dt$. The kinetic part of Eq. (9.64) leads to a white noise functional which is a two-dimensional version of Eq. (9.46). We have,

$$I_{xy} = N \exp\left\{\left(\frac{i+1}{2}\right) \int_0^T \left[\omega_x(t)^2 + \omega_y(t)^2\right] dt\right\}. \qquad (9.67)$$

On the other hand, the interaction part in Eq. (9.64) can be written as,

$$\exp\left[i\gamma \int_0^T (x\dot{y} - y\dot{x})\, dt\right] = \exp\left\{i\gamma \int_0^T [x_0 + B_x(t)]\,\omega_y(t)\, dt\right.$$

$$\left. -i\gamma \int_0^T [y_0 + B_y(t)]\,\omega_x(t)\, dt\right\}$$

$$= \exp\left\{i\gamma \int_0^T [x_0\omega_y(t) - y_0\omega_x(t)]\, dt\right\}$$

$$\times \exp(2i\gamma S_T), \qquad (9.68)$$

where S_T, given by

$$S_T = \frac{1}{2} \int_0^T [B_x(t)\, dB_y(t) - B_y(t)\, dB_x(t)], \qquad (9.69)$$

recalls Lévy's stochastic area, Eq. (2.34). This may be handled by noting that the two-dimensional Brownian motion, $B_x(t)$ and $B_y(t)$, can be realized on the probability space of a one-dimensional white noise where [61; 104; 142],

$$B_x(t) = \int_0^t \omega(\tau)\, d\tau \qquad ; \qquad dB_x(t) = \omega(t)\, dt \qquad (9.70)$$

$$B_y(t) = \int_{-t}^0 \omega(\tau)\, d\tau \qquad ; \qquad dB_y(t) = \omega(-t)\, dt\,. \qquad (9.71)$$

Equations (9.70) and (9.71) enable us to write Eq. (9.69), as (see Section 2.5),

$$S_T = \int_{\mathbf{R}^2} \omega(\tau_1)\, F_S(\tau_1, \tau_2)\, \omega(\tau_2)\, d\tau_1 d\tau_2$$

$$= \langle \omega, F_S(\tau_1, \tau_2)\, \omega\rangle \qquad (9.72)$$

where

$$F_S\left(\tau_1,\tau_2\right) = \frac{1}{4}\left[\chi_{[-T,0]}\left(\tau_1\right)\chi_{[0,-\tau_1]}\left(\tau_2\right) + \chi_{[-T,0]}\left(\tau_2\right)\chi_{[0,-\tau_2]}\left(\tau_1\right)\right]$$

$$-\frac{1}{4}\left[\chi_{[0,T]}\left(\tau_1\right)\chi_{[-\tau_1,0]}\left(\tau_2\right) + \chi_{[0,T]}\left(\tau_2\right)\chi_{[-\tau_2,0]}\left(\tau_1\right)\right].$$

$$(9.73)$$

The $\chi_{[\alpha,\beta]}\left(\tau_j\right)$, $j = 1,2$, with $[\alpha,\beta]$ as the limits of integration, denotes the integration over τ_j in Eq. (9.72).

With Eqs. (9.71) and (9.70), the terms in the exponent of the kinetic part, Eq. (9.67), can be written as,

$$\left(\frac{i+1}{2}\right)\int_0^T\left[\omega_x\left(\tau\right)^2 + \omega_y\left(\tau\right)^2\right]\,d\tau = \left(\frac{i+1}{2}\right)\int_0^T\left[\omega\left(\tau\right)^2 + \omega\left(-\tau\right)^2\right]\,d\tau$$

$$= -\frac{1}{2}\int_{-T}^T\omega\left(\tau\right)\,K\,\omega\left(\tau\right)\,d\tau, \qquad (9.74)$$

where $K = -\left(i+1\right)$. Furthermore, we again use the Donsker delta function $\delta\left(x\left(T\right) - x_1\right)$ and $\delta\left(y\left(T\right) - y_1\right)$ to fix the endpoints x_1 and y_1, where $x\left(T\right)$ and $y\left(T\right)$ are given by Eqs. (9.65) and (9.66). With Eqs. (9.68) to (9.74), we can now write Eq. (9.64) as,

$$K\left(x_1,y_1;x_0,y_0\right) = \int I_{xy}\exp\left[i\gamma\left(x_0\int_{-T}^0\omega\left(t\right)\,dt - y_0\int_0^T\omega\left(t\right)\,dt\right)\right]$$

$$\times\exp\left(-\frac{1}{2}\left\langle\omega,L\,\omega\right\rangle\right)$$

$$\times\delta\left(x_0 - x_1 + \int_0^T\omega\left(t\right)\,dt\right)$$

$$\times\delta\left(y_0 - y_1 + \int_{-T}^0\omega\left(t\right)\,dt\right)d\mu\left(\omega\right) \qquad (9.75)$$

where $L = -4i\gamma F_S\left(\tau_1,\tau_2\right)$. To evaluate the integration over the white noise measure, we write the delta functions in terms of their Fourier representation so that Eq. (9.75) becomes,

$$K\left(x_1, y_1; x_0, y_0\right) = \frac{1}{(2\pi)^2} \int\limits_{\mathbf{R}^2} d^2\mathbf{p} \exp\left[i\mathbf{p}\cdot(\mathbf{r}_0 - \mathbf{r}_1)\right]$$

$$\times \int N \exp\left[i\left(p_x - \gamma y_0\right) \int\limits_0^T \omega\left(t\right)\, dt\right]$$

$$\times \exp\left[i\left(p_y + \gamma x_0\right) \int\limits_{-T}^0 \omega\left(t\right) dt\right]$$

$$\times \exp\left(-\frac{1}{2}\langle\omega, K\omega\rangle\right) \exp\left(-\frac{1}{2}\langle\omega, L\,\omega\rangle\right)\, d\mu\left(\omega\right)$$

$$= \frac{1}{(2\pi)^2} \int\limits_{\mathbf{R}^2} d^2\mathbf{p} \exp\left[i\mathbf{p}\cdot(\mathbf{r}_0 - \mathbf{r}_1)\right] \int N \exp\left(i\langle\omega, \xi\rangle\right)$$

$$\times \exp\left(-\frac{1}{2}\langle\omega, K\omega\rangle\right) \exp\left(-\frac{1}{2}\langle\omega, L\,\omega\rangle\right)\, d\mu\left(\omega\right),$$

$$(9.76)$$

where $\xi = (p_x - \gamma y_0)\,\chi_{[0,T]} + (p_y + \gamma x_0)\,\chi_{[-T,0]}$, $\mathbf{r} = (x, y)$, and $\mathbf{p} = (p_x, p_y)$. We then note that the integration over $d\mu\left(\omega\right)$ is just the T-transform of the white noise functional, $\Phi = N\exp\left(-\frac{1}{2}\langle\omega, K\omega\rangle\right)\exp\left(-\frac{1}{2}\langle\omega, L\,\omega\rangle\right)$, i.e.,

$$(T\Phi)\left(\xi\right) = \int N \exp\left(i\langle\omega, \xi\rangle\right)\exp\left(-\frac{1}{2}\langle\omega, K\omega\rangle\right)\exp\left(-\frac{1}{2}\langle\omega, L\omega\rangle\right) d\mu\left(\omega\right)$$

$$= \left[\det\left(1 + L\left(1 + K\right)^{-1}\right)\right]^{-\frac{1}{2}}\exp\left[-\frac{1}{2}\left\langle\xi, (1 + K + L)^{-1}\,\xi\right\rangle\right],$$

$$(9.77)$$

where we used Eq. (33) of [61]. With $K = -\left(i + 1\right)$, we have,

$$\left[\det\left(1 + L\left(1 + K\right)^{-1}\right)\right]^{-\frac{1}{2}} = \left[\det\left(1 + iL\right)\right]^{-\frac{1}{2}}$$

$$= \left[\cos\left(\gamma T\right)\right]^{-1}, \qquad (9.78)$$

where we used Eq. (2.43). We also have [61],

$$\left\langle\xi, (1 + K + L)^{-1}\,\xi\right\rangle = (i/\gamma)\tan\left(\gamma T\right)\left[\left(p_x^2 + p_y^2\right)\right.$$

$$\left. + \gamma^2\left(x_0^2 + y_0^2\right) + 2\gamma\left(p_y x_0 - p_x y_0\right)\right]. \quad (9.79)$$

With Eqs. (9.77) to (9.79), Eq. (9.76) becomes,

$$K(x_1, y_1; x_0, y_0) = \frac{\cos(\gamma T)^{-1}}{(2\pi)^2} \exp\left[-\frac{i}{2}\gamma \tan(\gamma T)\left(x_0^2 + y_0^2\right)\right]$$

$$\times \int_{-\infty}^{+\infty} dp_x \exp\left\{-\left[i\tan(\gamma T)/2\gamma\right] p_x^2\right\}$$

$$\times \exp\left\{i\left[(x_0 - x_1) + \tan(\gamma T) y_0\right] p_x\right\}$$

$$\times \int_{-\infty}^{+\infty} dp_y \exp\left\{-\left[i\tan(\gamma T)/2\gamma\right] p_y^2\right\}$$

$$\times \exp\left\{i\left[(y_0 - y_1) - \tan(\gamma T) x_0\right] p_y\right\}. \qquad (9.80)$$

The Gaussian integration over dp_x and dp_y can be performed and we obtain,

$$K(x_1, y_1; x_0, y_0) = \frac{\gamma}{2\pi i \sin(\gamma T)} \exp\left\{\frac{i}{2}\gamma \cot(\gamma T)\left[(x_0 - x_1)^2 + (y_0 - y_1)^2\right]\right\}$$

$$\times \exp\left\{i\gamma(x_0 y_1 - x_1 y_0)\right\}. \qquad (9.81)$$

In terms of the Hermite polynomials H_n, this becomes,

$$K(x_1, y_1; x_0, y_0) = (\gamma/\pi) \exp\left\{i\gamma(x_0 y_1 - x_1 y_0)\right\}$$

$$\times \exp\left\{(-\gamma/2)\left[(x_1 - x_0)^2 + (y_1 - y_0)^2\right]\right\}$$

$$\times \sum_{n=0}^{\infty}\sum_{m=0}^{n} \frac{(-1)^{-n} 2^{-2n}}{m!(n-m)!}$$

$$\times H_{2m}\left(\sqrt{\gamma}(x_1 - x_0)\right) H_{2(n-m)}\left(\sqrt{\gamma}(y_1 - y_0)\right)$$

$$\times \exp\left[-i\left(n + \frac{1}{2}\right) 2\gamma T\right]. \qquad (9.82)$$

From the propagator, the energy spectrum of the particle is obtained as $E = \left(n + \frac{1}{2}\right) e\mathcal{B}$.

9.7   Outlook

Sections 9.5 and 9.6 illustrate that one can use the infinite-dimensional white noise analysis to evaluate the path integral for nonrelativistic quantum mechanical systems. By parametrizing the paths in terms of the white noise variable $\omega(\tau)$ and expressing the Feynman integrand as a white noise functional, an integration over the Gaussian white noise measure $d\mu(\omega)$ allows us to obtain the quantum propagator for different systems.

The applications of white noise path integrals in quantum mechanics can, however, be extended in several directions. For example, certain quantum systems are solvable only when expressed in curvilinear coordinates (see, e.g., [49–52]), as exemplified by the Coulomb problem. Although one can use the present approach to treat angular variables (e.g., particle in a circle [21; 140]), the proper handling of the radial variable $(0 \leq r < \infty)$ has yet to be done. Other issues may come to the fore in pursuing these directions, such as the use of time rescaling (to solve the Coulomb problem [107; 129]) in the context of white noise analysis.

Exercises

(9-1) From the Schrödinger equation,

$$\left[\frac{-\hbar^2 \boldsymbol{\nabla}^2}{2\mu} + V(\mathbf{r})\right] \Psi(\mathbf{r}, t) = i\hbar \frac{\partial}{\partial t} \Psi(\mathbf{r}, t)$$

and

$$\Theta(t' - t)\, \Psi(\mathbf{r}', t') = \int K(\mathbf{r}', t'; \mathbf{r}, t)\, \Psi(\mathbf{r}, t)\, d^3\mathbf{r},$$

obtain the differential equation satisfied by the propagator $K(\mathbf{r}', t'; \mathbf{r}, t)$. The $\Theta(\tau)$ is a unit step function $(\Theta(\tau) = 1,$ for $\tau > 0,$ and $\Theta(\tau) = 0,$ for $\tau < 0)$ which assures the forward propagation of Ψ in time $(t' > t)$. Note that $d\Theta(\tau)/d\tau = \delta(\tau)$.

(9-2) Consider the propagator, $K(\mathbf{r} + \boldsymbol{\eta},\ t + \Delta t;\ \mathbf{r}, t)$, which satisfies the Schrödinger equation,

$$\left[\frac{-\hbar^2 \boldsymbol{\nabla}^2}{2\mu} + V(\mathbf{r})\right] K = i\hbar \frac{\partial K}{\partial t},$$

where $\boldsymbol{\eta} = (x, y, z)$ and Δt are infinitesimal. Show that by expanding K as,

$$K(\mathbf{r} + \boldsymbol{\eta},\ t + \Delta t;\ \mathbf{r}, t) = K + \frac{\partial K}{\partial t}\Delta t + \cdots$$

one obtains the short-time propagator,

$$K\left(\mathbf{r}+\boldsymbol{\eta},\ t+\Delta t;\ \mathbf{r},t\right) \cong \left(\frac{\mu}{2\pi i\hbar\Delta t}\right)^{3/2}$$

$$\times \exp\left\{\frac{i}{\hbar}\left[\frac{1}{2}\mu\left(\frac{\boldsymbol{\eta}}{\Delta t}\right)^2 - V\left(\mathbf{r}\right)\right]\Delta t\right\},$$

where $\lim\limits_{\Delta t\to 0} K\left(\mathbf{r}+\boldsymbol{\eta};\ t+\Delta t;\ \mathbf{r},t\right) = \delta\left(\boldsymbol{\eta}\right)$.

(9-3) Using the white noise functional approach discussed in Section 9.5, consider a particle in the potential $V = kx$, and obtain the quantum propagator for a particle subjected to a constant force given by,

$$K(x'',x';t) = \sqrt{\frac{\mu}{2\pi i\hbar t}}\, \exp\left[\frac{i\mu}{2\hbar t}\left(x''-x'\right)^2 - \frac{i}{2\hbar}\left(x''+x'\right)kt - \frac{i}{24\mu\hbar}(k^2t^3)\right].$$

(9-4) Show that Eq. (9.81) can be written as Eq. (9.82) in terms of the Hermite polynomials.

Chapter 10

Quantum Particles with Boundary Conditions

Rapidly advancing high technology instrumentation and development of new materials are giving deeper insight into interactions between quantum systems with boundaries. Often, stochasticity at varying scales appears as a compounding factor in the analysis and modelling of such systems. We are, therefore, motivated to see how we can have a white noise analysis of systems with nontrivial boundaries.

In this chapter, we look at the problem of a quantum particle moving in a region with periodic boundaries, in particular, circular topologies including the Aharonov-Bohm setup. We also look at the infinite wall potential and combinations thereof, in particular, the box potential. Furthermore, we also treat the free particle on the half-line with general boundary conditions. This would include Dirichlet and Neumann boundary conditions as special cases.

10.1 Quantum Particle with Periodic Boundaries

In 1967, a pioneering work of S.F. Edwards [77] tackled topological constraints in statistical mechanics in the context of polymer entanglements using the correspondence between the Feynman path integral and the differential Schrödinger equation. In particular, the topological constraints investigated involve a point singularity in a plane and a ring singularity in three dimensions [77; 179] which were later used to study the Aharonov-Bohm experiment, and other interesting applications in condensed matter physics and cosmology. We now look at such systems from the perspective of white noise path integrals.

In handling topological defects, such as a punctured plane [31; 35; 36; 76; 77; 91; 137; 185] where the hole makes the physical space multiply-connected, the Feynman path integral approach appears to be the most suitable. The approach, being a sum over all possible histories of the particle, accounts for the presence of the topological defect by classifying the different particle trajectories into homotopically inequivalent paths characterized by a winding number n. A nice simple example of a space which is multiply connected is a circle. We first encountered this topology in our earlier discussion for helical biopolymers where the solution for the corresponding Fokker-Planck equation is obtained as a white noise path integral. We now consider the quantum case where a particle is constrained to move in a circle. The results can be applied to a number of quantum systems that exhibit periodic boundary conditions as exemplified by an electron in a periodic latttice.

We now consider the motion of a particle which starts at a point φ_0 in a circle and ends at φ_1. In going from φ_0 to φ_1, the particle may move in the clockwise or anti-clockwise direction. In fact, many inequivalent paths can be generated if the particle first winds several times, clockwise or anti-clockwise, before ending at φ_1. The paths can then be classified by winding numbers n which signifies, if n is positive, an n-times counterclockwise path around the circle, and if negative, an $|n| - 1$ times clockwise winding (see Figure 3.2).

In solving for the quantum particle propagator $K(\varphi_1, \varphi_0; \tau)$, we can sum over all possible topologically inequivalent paths and evaluate [36; 137; 185],

$$K(\varphi_1, \varphi_0; \tau) = \sum_{n=-\infty}^{+\infty} K_n(\varphi_1, \varphi_0; \tau), \qquad (10.1)$$

where $K_n(\varphi_1, \varphi_0; \tau)$, is the partial propagator for particle paths belonging to a winding number n. The procedure that can be followed, therefore, is to calculate for $K_n(\varphi_1, \varphi_0; \tau)$ and then sum it for all values of n. We first calculate the propagator for a winding number $n = 0$ by considering the path integral ($\hbar = 1$),

$$K_0(\varphi_1, \varphi_0; \tau) = \int \exp[iS] \; \mathcal{D}[\varphi], \qquad (10.2)$$

with the action,

$$S = \int_0^\tau \left(\frac{1}{2}I\dot{\varphi}^2\right) dt, \qquad (10.3)$$

where $I = MR^2$, for a particle of mass M moving in a circle of radius R. The propagator, Eq. (10.2), can be evaluated as a white noise path integral. We parametrize the paths of a particle which start from a point φ_0 as,

$$\varphi(\tau) = \varphi_0 + \left(1/\sqrt{I}\right) B\left(\tau\right)$$

$$= \varphi_0 + \left(1/\sqrt{I}\right) \int_0^\tau \omega\left(t\right) dt \tag{10.4}$$

where $B\left(t\right)$ is ordinary Brownian motion and $\omega\left(\tau\right)$ is the white noise variable. The time derivative of Eq. (10.4) is simply, $\dot{\varphi} = \left(1/\sqrt{I}\right)\omega$ and the exponential in Eq. (10.2) can be written as,

$$\exp\left(iS\right) = \exp\left[i\int_0^\tau \left(\frac{1}{2}I\dot{\varphi}^2\right) dt\right]$$

$$= \exp\left[\frac{i}{2}\int_0^\tau \omega(t)^2\, dt\right]. \tag{10.5}$$

Having expressed $\exp[iS]$ in terms of $\omega(t)$, the integration over all paths with $\mathcal{D}\left[\varphi\right]$ leads to an integration over the Gaussian white noise measure $d\mu(\omega)$. We can then apply the white noise procedure. In particular, we have the correspondence, $\mathcal{D}\left[\varphi\right] \to N_\omega\, d^\infty\omega$. From Eq. (2.7) we have, $N_\omega\, d^\infty\omega = \exp\left(\frac{1}{2}\int \omega(t)^2\, dt\right) d\mu(\omega)$, and the exponential factor $\exp\left(\frac{1}{2}\int \omega(t)^2\, dt\right)$ combines with $\exp(iS)$ given by Eq. (10.5) to yield the Gauss kernel (see also Eq. (9.46)),

$$I_0 = \mathcal{N}\ \exp\left[\left(\frac{i+1}{2}\right)\int_0^\tau \omega(t)^2\, dt\right], \tag{10.6}$$

where $\mathcal{N}$ is a normalization. Although the path parametrization Eq. (10.4) fixes the initial point φ_0, the paths belonging to $n = 0$ still need to be pinned down at the final point φ_1. This is done using the Donsker delta function, $\delta\left(\varphi(t) - \varphi_1\right)$. This, together with Eq. (10.6), allows us to write the propagator Eq. (10.2), as

$$K_0\left(\varphi_1, \varphi_0; \tau\right) = \int I_0\ \delta\left(\varphi(t) - \varphi_1\right)\ d\mu(\omega). \tag{10.7}$$

This equation for winding number $n = 0$ can be generalized for paths belonging to winding number $n \neq 0$ which are likewise pinned down at φ_1 by using the delta function, $\delta\left(\varphi(t) - \varphi_1 + 2\pi n\right)$ where $n = 0, \pm, 1, \pm 2, ...$, i.e.,

$$K_n\left(\varphi_1, \varphi_0; \tau\right) = \int I_0 \, \delta\left(\varphi(t) - \varphi_1 + 2\pi n\right) \, d\mu(\omega). \qquad (10.8)$$

Summing over all possible paths, the full quantum propagator Eq. (10.1) can now be expressed as,

$$K\left(\varphi_1, \varphi_0; \tau\right) = \sum_{n=-\infty}^{+\infty} \int I_0 \, \delta\left(\varphi(t) - \varphi_1 + 2\pi n\right) d\mu(\omega)$$

$$= \sum_{n=-\infty}^{+\infty} \int I_0 \delta\left(\varphi_0 + \frac{1}{\sqrt{I}}\int_0^\tau \omega(t) \, dt - \varphi_1 + 2\pi n\right) d\mu(\omega),$$

$$(10.9)$$

where we used Eq. (10.4). Using the Poisson sum formula, Eq. (8.13), we can rewrite Eq. (10.9) as,

$$K\left(\varphi_1, \varphi_0; \tau\right) = \frac{1}{2\pi} \sum_{m=-\infty}^{+\infty} \int I_0 \exp\left[im\left(\varphi_0 + \frac{1}{\sqrt{I}}\int_0^\tau \omega(t) \, dt - \varphi_1\right)\right] d\mu(\omega)$$

$$= \frac{1}{2\pi} \sum_{m=-\infty}^{+\infty} \exp\left[im\left(\varphi_0 - \varphi_1\right)\right]$$

$$\times \int I_0 \exp\left[i\left(\int_0^\tau \omega \xi \, d\tau\right)\right] d\mu(\omega), \qquad (10.10)$$

where $\xi = m/\sqrt{I}$. Finally, the integration over the white noise measure $d\mu(\omega)$ is just the T-transform of I_0, Eq. (2.28). We thus get,

$$K\left(\varphi_1, \varphi_0; \tau\right) = \frac{1}{2\pi} \sum_{m=-\infty}^{+\infty} \exp\left[im\left(\varphi_0 - \varphi_1\right)\right] \exp\left(-\frac{i}{2}\int_0^\tau \xi^2 dt\right)$$

$$= \frac{1}{2\pi} \sum_{m=-\infty}^{+\infty} \exp\left[im\left(\varphi_0 - \varphi_1\right)\right] \exp\left(-iE_m \tau\right), \qquad (10.11)$$

where the energy spectrum is $E_m = m^2/2MR^2$, with m as the angular quantum number.

10.2 The Aharonov-Bohm Setup

The Aharonov-Bohm effect gave rise to much controversy when it was first brought to light [4]. Classically, since the electron moves in a space where the magnetic field $\mathbb{B}$ is zero (outside an impenetrable solenoid), then the force on the electron due to $\mathbb{B}$ is also zero, i.e., force $\mathbb{F} = e\mathbf{v} \times \mathbb{B} = 0$, where $\mathbf{v}$ is the velocity of the electron. It was argued, for example, that changes in the value of the magnetic field $\mathbb{B}$ inside the solenoid should not affect the electron. The result, however, shows that changes in $\mathbb{B}$ do affect, for instance, the electron's energy E_m^Φ which depends on the value of the magnetic flux $\Phi = \pi r_0{}^2 B$, where B is the magnitude of $\mathbb{B}$. Successive experimentation did confirm the existence of the Aharonov-Bohm effect. For example, an impenetrable toroid containing a magnetic flux was used to show that electrons passing outside the toroidal ring and through the hole of the ring exhibit interference patterns which are affected by changes in values of the magnetic flux, even if the electrons are not in contact with $\mathbb{B}$ [170].

Consider an impenetrable solenoid of radius r_0, oriented along the z-axis, carrying a magnetic flux Φ such that, outside the solenoid, the magnetic field is zero, i.e., $\mathbb{B} = \nabla \times \mathbf{A} = 0$. The vector potential $\mathbf{A}$ for this configuration is given by,

$$\mathbf{A} = \frac{\Phi}{2\pi} \left(\frac{-y\hat{i} + x\hat{j}}{x^2 + y^2} \right) \quad ; \quad x^2 + y^2 > r_0 \,. \tag{10.12}$$

The Aharonov-Bohm setup has a charged particle moving in the space outside the solenoid with the confined magnetic flux [4]. We can take advantage of the cylindrical symmetry of the problem by looking at the cross-section, or the $(x - y)$-plane, to write the action as,

$$S^{AB} = \int \left[\frac{1}{2} M \dot{\mathbf{r}}^2 + \left(\frac{e}{c} \right) \mathbf{A} \cdot \dot{\mathbf{r}} \right] dt \,, \tag{10.13}$$

where $\mathbf{r} = (x, y)$. The charged particle may further be constrained to move in a circle of radius R perpendicular to the z-axis with the solenoid at the center coinciding with the origin. In polar coordinates, $\mathbf{r} = (r, \varphi)$, the vector potential can be written as,

$$\mathbf{A} = (\Phi/2\pi) \, \boldsymbol{\nabla}\varphi \,, \qquad (r > r_0) \,, \qquad (10.14)$$

and the action, Eq. (10.13), acquires the form,

$$S^{AB} = \int \left[\frac{1}{2}I\dot{\varphi}^2 + \left(\frac{e\Phi}{2\pi c} \right)\dot{\varphi} \right] dt \,, \qquad (10.15)$$

where $I = MR^2$. Note that the only difference between Eq. (10.15) and Eq. (10.3) is the presence of the potential, $(e\Phi/2\pi c)\,\dot{\varphi}$. This problem can be cast in the language of white noise by parametrizing the paths of the particle using Eq. (10.4), where Eq. (10.15) becomes,

$$S^{AB} = \int \left[\frac{1}{2}\omega^2 + \left(\frac{e\Phi}{2\pi c\sqrt{I}} \right)\omega \right] dt \,. \qquad (10.16)$$

Following the steps in the previous section, the quantum propagator for this problem takes the form,

$$K^{AB}(\varphi_1, \varphi_0; \tau) = \sum_{n=-\infty}^{+\infty} \int I_0 \, \delta\left(\varphi(t) - \varphi_1 + 2\pi n\right) \exp\left[\frac{i\alpha}{\sqrt{I}} \int_0^\tau \omega\,(t)\, dt \right] d\mu(\omega)$$

$$= \sum_{n=-\infty}^{+\infty} \int I_0 \, \delta\left(\varphi_0 + \frac{1}{\sqrt{I}} \int_0^\tau \omega\,(t)\, dt - \varphi_1 + 2\pi n \right)$$

$$\times \exp\left[\frac{i\alpha}{\sqrt{I}} \int_0^\tau \omega\,(t)\, dt \right] d\mu(\omega) \,, \qquad (10.17)$$

where $\alpha = e\Phi/2\pi c$. Note that Eq. (10.17) differs from Eq. (10.9) in view of the potential, $\left(i\alpha/\sqrt{I} \right) \int \omega\,(t)\; dt$. Utilizing the Poisson sum formula,

Eq. (8.13), the propagator becomes,

$$K(\varphi_1, \varphi_0; \tau) = \frac{1}{2\pi} \sum_{m=-\infty}^{+\infty} \int I_0 \exp\left[im\left(\varphi_0 + \frac{1}{\sqrt{I}}\int_0^\tau \omega(t)\,dt - \varphi_1\right)\right]$$

$$\times \exp\left[\frac{i\alpha}{\sqrt{I}}\int_0^\tau \omega(t)\,dt\right] d\mu(\omega)$$

$$= \frac{1}{2\pi} \sum_{m=-\infty}^{+\infty} \exp\left[im(\varphi_0 - \varphi_1)\right]$$

$$\times \int I_0 \exp\left[i\left(\frac{m+\alpha}{\sqrt{I}}\right)\int_0^\tau \omega(t)\,dt\right] d\mu(\omega)$$

$$= \frac{1}{2\pi} \sum_{m=-\infty}^{+\infty} \exp[im(\varphi_0 - \varphi_1)] \int I_0 \exp\left[i\int_0^\tau \omega(t)\,\xi_\alpha dt\right] d\mu(\omega),$$

$$(10.18)$$

where $\xi_\alpha = (m+\alpha)/\sqrt{I}$. The integration over $d\mu(\omega)$ is again the T-transform of I_0 (see Eq. (2.28)), and Eq. (10.18) reduces to,

$$K(\varphi_1, \varphi_0; \tau) = \frac{1}{2\pi} \sum_{m=-\infty}^{+\infty} \exp\left[im(\varphi_0 - \varphi_1)\right] \exp\left(-\frac{i}{2}\int_0^\tau \xi_\alpha^2 dt\right)$$

$$= \frac{1}{2\pi} \sum_{m=-\infty}^{+\infty} \exp\left[im(\varphi_0 - \varphi_1) - iE_m^\Phi \tau\right]. \qquad (10.19)$$

Here, the energy spectrum is $E_m^\Phi = [m + (e\Phi/2\pi c)]^2 / 2MR^2$. Compared with the previous section, the angular quantum number m is modified by the magnetic flux $e\Phi/2\pi c$. For the case, $\Phi = 0$, the Aharonov-Bohm propagator reduces to that of the particle in a circle.

10.3   Infinite Wall Potential

As preparation for discussing general boundary conditions, we first consider an example of a quantum particle in a space with flat wall boundary [21]. Specifically, we take the potential which describes an infinite wall at the origin, i.e.,

$$V(x) = \begin{cases} \infty, & \text{for} \quad x \le 0 \\ 0, & \text{for} \quad x > 0 \,. \end{cases} \qquad (10.20)$$

Classically, for a particle of mass M which goes from x_0 to x_1 there are two possible paths: the first is a path that goes directly from the initial to the final point, and the second describes a path from x_0 that is reflected by the wall before arriving at x_1. Quantum mechanically, the particle propagator satisfies the boundary condition,

$$K(x_1, x_0) = 0, \quad \text{at} \quad x_1 = 0 \text{ or } x_0 = 0 \,, \qquad (10.21)$$

together with,

$$\lim_{\tau \to 0} K(x_1, x_0; \tau) = \delta(x_1 - x_0) \,. \qquad (10.22)$$

For this problem we start with the Gauss kernel I_0, Eq. (9.46), together with the Donsker delta function, $\delta\left(x(\tau) - x_1\right)$, and write the linear combination of white noise functionals ($\hbar = 1$),

$$I^W(x_1, \tau \mid x_0, 0) = I_0 \, \delta\left(x_0 + (1/M)^{1/2} \int_0^\tau \omega(t) \, dt - x_1\right)$$

$$- I_0 \, \delta\left(-x_0 + (1/M)^{1/2} \int_0^\tau \omega(t) \, dt - x_1\right). \qquad (10.23)$$

The two terms on the right-hand side are Feynman integrands of the form given by Eq. (9.47). The second term arises from particle trajectories originating from an image point, $-x_0$, and arriving at x_1. At the origin $x_0 = 0$ where the infinite wall is located, the two terms cancel out and the $I^W(x_1, \tau \mid x_0, 0)$ vanishes. For a propagator satisfying the boundary condition, Eq. (10.21), the combination of the type given by Eq. (10.23) has been discussed in the literature [119; 185].

Writing the delta function in terms of its Fourier representation we express the functional in Eq. (10.23) as,

$$I^W(x_1, \tau \mid x_0, 0) = \frac{1}{2\pi} \int_{-\infty}^{+\infty} dk \, \{\exp[ik(x_0 - x_1)] - \exp[-ik(x_0 + x_1)]\}$$

$$\times I_0 \, \exp\left[(ik/\sqrt{M}) \int_0^\tau \omega(t) \, dt\right]. \qquad (10.24)$$

We then integrate Eq. (10.24) over the Gaussian white noise measure $d\mu(\omega)$ to obtain the propagator,

$$K^W(x_1, x_0; \tau) = \int I^W(x_1, \tau \mid x_0, 0) \, d\mu(\omega)$$

$$= \frac{1}{2\pi} \int\limits_{-\infty}^{+\infty} dk \, \{\exp[ik(x_0 - x_1)] - \exp[-ik(x_0 + x_1)]\}$$

$$\times \int I_0 \exp\left[(ik/\sqrt{M}) \int\limits_0^\tau \omega(t) \, dt\right] d\mu(\omega). \quad (10.25)$$

The integral for the Gaussian white noise is just the T-transform of the Gauss kernel I_0 (see Eqs. (2.8) and (2.28)), with $\xi = k/\sqrt{M}$. Hence, the quantum propagator for the infinite wall potential becomes,

$$K^W(x_1, x_0; \tau) = \frac{1}{2\pi} \int\limits_{-\infty}^{+\infty} \{\exp[ik(x_0 - x_1)] - \exp[-ik(x_0 + x_1)]\}$$

$$\times \exp\left(-\frac{i}{2} \int\limits_0^\tau (k/\sqrt{M})^2 dt\right) dk, \quad (10.26)$$

which could be written as,

$$K^W(x_1, x_0; \tau) = \int\limits_{-\infty}^{+\infty} \Psi_k(x_0) \, \Psi_k(x_1) \exp(-iE_k\tau) \, dk, \quad (10.27)$$

where the energy is, $E_k = (k^2/2M)$, and the eigenfunction is given by, $\Psi_k(x) = (1/\sqrt{\pi}) \sin(kx)$.

10.4 Particle in a Box

Our second application treats the standard problem of a particle of mass M in a one-dimensional box of length L, with sides located at $x = 0$ and $x = L$. Classically, the paths of the particle in a box can be categorized into four classes [185]. The first goes directly from the initial point x_0 to the final point x_1 without hitting the walls; the second type of path hits the wall at $x = 0$ before arriving at x_1; the third leaves x_0 and is reflected from the wall at $x = L$ before reaching x_1; and the fourth class describes a path which bounces once from the boundary at $x = 0$ and also once from the wall at $x = L$ before reaching x_1. The other paths belonging to any of the four classes describe a particle bouncing back and forth inside the box of length L and, therefore, travelling an additional distance of $2Ln$ $(n = 0, 1, 2, ...)$.

Quantum mechanically, the particle propagator $K(x_1, x_0; t)$ has to satisfy the boundary conditions,

$$K(x_1, 0; t) = K(x_1, L; t) = K(0, x_0; t) = K(L, x_0; t) = 0, \qquad (10.28)$$

aside from the requirement,

$$\lim_{t \to 0} K(x_1, x_0; t) = \delta(x_1 - x_0). \qquad (10.29)$$

To incorporate all the possible paths of the particle, we use the form of the Feynman integrand for a free particle, Eq. (9.47), and write the white noise functional for a particle in a box as,

$$I^B(x_1, t \mid x_0, 0) = \sum_{n=-\infty}^{+\infty} I_0 \, C_n(x_1, t \mid x_0, 0), \qquad (10.30)$$

where [112],

$$2C_n(x_1, t \mid x_0, 0) = \delta\left(x(t) - x_1 + 2Ln\right) + \delta\left(-x(t) + x_1 + 2Ln\right)$$
$$- \delta\left(x(t) + x_1 + 2Ln\right) - \delta\left(-x(t) - x_1 + 2Ln\right), \qquad (10.31)$$

with $x(t)$ given by Eq. (9.44) with $\hbar = 1$. The $(2Ln)$ in the delta function describes the fact that the particle leaving x_0 can bounce back and forth inside the box of length L before arriving at x_1. Combining the first two terms, as well as the last two terms in Eq. (10.31), we write Eq. (10.30) as,

$$I^B(x_1, t \mid x_0, 0) = \sum_{n=-\infty}^{+\infty} I_0 \left\{ \delta\left(x_0 + (1/M)^{1/2} \int_0^t \omega(\tau) \, d\tau - x_1 + 2Ln\right) \right.$$
$$\left. - \delta\left(x_0 + (1/M)^{1/2} \int_0^t \omega(\tau) \, d\tau + x_1 + 2Ln\right) \right\}. \qquad (10.32)$$

Writing $\delta\left(x(t) - x_1 + 2Ln\right) = (\pi/L) \, \delta\left((\pi/L)[x(t) - x_1] + 2\pi n\right)$, we apply the Poisson sum formula [17],

$$\sum_{n=-\infty}^{\infty} \delta(\phi + 2\pi n) = (1/2\pi) \sum_{m=-\infty}^{\infty} \exp\left(im\phi\right), \qquad (10.33)$$

and Eq. (10.32) becomes,

$$I^B(x_1, t \mid x_0, 0) = 1/2L \sum_{m=-\infty}^{+\infty} I_0 \exp\left[\left(im\pi/L\sqrt{M}\right) \int_0^t \omega(\tau)\, d\tau\right]$$

$$\times \left\{ \exp\left[(im\pi/L)(x_0 - x_1)\right] - \exp\left[(im\pi/L)(x_0 + x_1)\right] \right\}. \tag{10.34}$$

To obtain the quantum propagator we integrate $I^B(x_1, t \mid x_0, 0)$ over the Gaussian white noise measure $d\mu(\omega)$. This integral is just the T-transform of I_0, Eq. (2.28), where $\xi = m\pi/L\sqrt{\mu}$, i.e.,

$$K^B(x_1, x_0; t) = (1/2L) \sum_{m=-\infty}^{+\infty} \int I_0 \exp\left[\left(im\pi/L\sqrt{M}\right) \int_0^t \omega(\tau)\, d\tau\right] d\mu(\omega)$$

$$\times \left\{ \exp\left[(im\pi/L)(x_0 - x_1)\right] - \exp\left[(im\pi/L)(x_0 + x_1)\right] \right\}$$

$$= (1/2L) \sum_{m=-\infty}^{+\infty} \exp\left[-\frac{i}{2} \int_0^t \left(m\pi/L\sqrt{M}\right)^2 d\tau\right]$$

$$\times \left\{ \exp\left[(im\pi/L)(x_0 - x_1)\right] - \exp\left[(im\pi/L)(x_0 + x_1)\right] \right\}. \tag{10.35}$$

We can further express Eq. (10.35) as,

$$K^B(x_1, x_0; t) = (1/2L) \sum_{m=-\infty}^{+\infty} \exp\left[-(i/2)\left(m\pi/L\sqrt{M}\right)^2 t\right]$$

$$\times \left[\cos\left(\frac{m\pi}{L}(x_0 - x_1)\right) - \cos\left(\frac{m\pi}{L}(x_0 + x_1)\right)\right], \tag{10.36}$$

with the sum over terms such as $\sin\left(\frac{m\pi}{L}(x_0 - x_1)\right)$ giving a zero contribution. Equation (10.36) can be written in the symmetrized form,

$$K^B(x_1, x_0; t) = \sum_{m=-\infty}^{+\infty} \phi_m(x_0)\phi_m(x_1) \exp(-iE_m t), \tag{10.37}$$

where $E_m = m^2\pi^2/2ML^2$ and $\phi_m(x) = (1/\sqrt{L})\sin(m\pi x/L)$, are the energy eigenvalues and eigenfunctions for a quantum particle in a one-dimensional box.

10.5   Free Particle on the Half-line with General Boundary Conditions

In this section, we consider the white noise path integral treatment of a quantum propagator $K^{\beta/\alpha}(x,t \mid x_0, t_0)$ for the half-line with the general or Robin boundary condition of the form,

$$\left[\alpha\partial_x K^{\beta/\alpha}(x,t \mid x_0, t_0) - \beta\, K^{\beta/\alpha}(x,t \mid x_0, t_0)\right]_{x=0} = 0. \qquad (10.38)$$

Equation (10.38) is a combination of the Neumann condition satisfied for, $\beta = 0$, i.e., $\left[\partial_x K^0(x,t \mid x_0, t_0)\right]_{x=0} = 0$, and the Dirichlet boundary condition for $\alpha = 0$ or $\gamma = (\beta/\alpha) \to \infty$, i.e., $K^{\gamma\to\infty}(x,t \mid x_0, t_0)|_{x=0}= 0$. The general boundary condition at the origin gives rise to a Dirac δ-function potential, $V_\gamma(x) = \gamma\delta(x)$. This is discussed in the Feynman path integral treatments of Farhi and Gutmann [82] which analytically continues to Brownian motion path integrals, and Clark, Menikoff and Sharp [58] which derives the Feynman sum over histories as a continuum limit of a discrete random walk on the half-line with an elastic barrier at the origin. For $\gamma > 0$, a continuous spectrum is obtained, while a bound state arises from the reverse condition $\gamma < 0$. Physical applications of interest include sticky walls [195], mesic atoms with short-range interactions [6], permeable membranes, and walls with both absorption and reflection of wave functions [33].

Consider an interaction potential given by $V(x) = \gamma\,\delta(x)$. From Eq. (9.49) for a particle in this potential, and Eq. (10.23) for the boundary at $x = 0$, we have the Feynman integrand given by the white noise functional [25; 195],

$$I_\delta = I_0\left[\delta(x(t) - x) + \delta(x(t) + x)\right]e^{-\frac{i}{\hbar}\gamma \int \delta(x)d\tau}. \qquad (10.39)$$

Taking the T-transform of Eq. (10.39), i.e., $TI_\delta(0)$, yields the propagator,

$$
\begin{aligned}
K_\delta^{\beta/\alpha}(x,t \mid x_0, t_0) &= TI_\delta(0) \\
&= T\left(I_0\delta(x(t) - x)\,e^{-\frac{i}{\hbar}\int \gamma\,\delta(x)d\tau}\right)(0) \\
&\quad + T\left(I_0\delta(x(t) + x)\,e^{-\frac{i}{\hbar}\int \gamma\,\delta(x)d\tau}\right)(0) \\
&= K_\delta(x,t \mid x_0, t_0) + K_\delta(-x,t \mid x_0, t_0). \qquad (10.40)
\end{aligned}
$$

The propagators appearing in the linear combination, Eq. (10.40), can be obtained as a special case of the solution for the general class of potentials, $\widehat{V}(x) = \gamma\,\delta(x) + V(|x|)$. The usual Dyson series for the Feynman propagator is obtained as,

$$
K_V\left(x,t \mid x_0,t_0\right) = \sum_{n=0}^{\infty} K_n\left(x,t \mid x_0,t_0\right)
$$

$$
= K_0\left(x,t \mid x_0,t_0\right) + \sum_{n=1}^{\infty} (-i)^n \prod_{j=1}^{n} \int dx_j \int dt_j V\left(x_j\right)
$$

$$
\times \prod_{k=1}^{n+1} K_0\left(x_k,t_k \mid x_{k-1},t_{k-1}\right). \tag{10.41}
$$

The iterated integrals $\prod \int dt_j$ give a convolution of Laplace transforms. For the nth term, application of the Laplace transform [81] gives,

$$
\int_0^{\infty} e^{-pt}\left(\prod_{j=1}^{n} \int dt_j \prod_{k=1}^{n+1} K_0\left(x_k,t_k \mid x_{k-1},t_{k-1}\right) \right) dt
$$

$$
= (2ip)^{-\frac{n+1}{2}} \exp\left[-(1-i)\sqrt{\alpha_n}\sqrt{p}\right], \tag{10.42}
$$

where $\sqrt{\alpha_n} = \left(|x-x_n| + |x_n-x_{n-1}| + ... + |x_1-x_0|\right)$. The inverse Laplace transform of Eq. (10.42) yields for the nth term (see also, Eq. (7.728) in Gradshteyn and Ryzhik [98]) the result,

$$
\prod_{j=1}^{n} \int dt_j \prod_{k=1}^{n+1} K_0\left(x_k,t_k \mid x_{k-1},t_{k-1}\right) = \frac{1}{\sqrt{2\pi}} e^{-\frac{i\pi(n+1)}{4}} t^{\frac{n-1}{2}} e^{-\alpha_n/4it}
$$

$$
\times D_{-n}\left(\sqrt{\alpha_n/it}\right), \tag{10.43}
$$

where $D_n\left(\cdot\right)$ is the parabolic cylinder function. We, therefore, obtain for the propagator of the class of time-independent potentials $\widehat{V}\left(x\right)$ [25; 195],

$$
K_{\widehat{V}}\left(x,t \mid x_0,0\right) = K_0\left(x,t \mid x_0,0\right) + \frac{1}{\sqrt{2\pi}} \sum_{n=1}^{\infty} (-i)^{\frac{3n+1}{2}} t^{\frac{n-1}{2}} \prod_{j=1}^{n} \int dx_j \widehat{V}(x_j)
$$

$$
\times \exp\left(-\alpha_n\left(x;x_j,...\right)/4it\right) D_{-n}\left(\sqrt{\alpha_n\left(x;x_j,...\right)/it}\right), \tag{10.44}
$$

where $\alpha_n(x;x_j,...) = \left(|x-x_n| + |x_n-x_{n-1}| + ... + |x_1-x_0|\right)^2$ and $K_0(x,t \mid x_0,0)$ is the free particle propagator. Taking $V\left(|x|\right) = 0$ in $\widehat{V}\left(x\right)$

of Eq. (10.44), and integrating over dx_j we have,

$$K_\delta\left(x,t \mid x_0,0\right) = K_0\left(x,t \mid x_0,0\right)$$

$$+ \frac{1}{\sqrt{2\pi}} e^{-z^2/4} \sum_{n=1}^{\infty} \gamma^n\left(-i\right)^{\frac{3n+1}{2}} t^{\frac{n-1}{2}} D_{-n}\left(z\right), \quad (10.45)$$

where $z = \left(|x|+|x_0|\right)/\sqrt{it}$. Using the integral representation [98] of $D_{-n}\left(z\right)$ the sum in the second term can be expressed as,

$$\sum_{n=1}^{\infty} \gamma^n\left(-i\right)^{\frac{3n+1}{2}} t^{\frac{n-1}{2}} D_{-n}\left(z\right) = \sum_{n=1}^{\infty} \gamma^n\left(-i\right)^{\frac{3n+1}{2}} t^{\frac{n-1}{2}} \left(e^{-z^2/4}/\Gamma\left(n\right)\right)$$

$$\times \int_0^{\infty} e^{-zy-\left(y^2/2\right)} y^{n-1} dy. \quad (10.46)$$

In Eq. (10.46), the series in n is summable,

$$\sum_{n=1}^{\infty} \gamma^n\left(-i\right)^{\frac{3n+1}{2}} t^{\frac{n-1}{2}} \left(y^{n-1}/\Gamma\left(n\right)\right) = \gamma e^{-\gamma\sqrt{it}\, y}. \quad (10.47)$$

With Eqs. (10.46) and (10.47), Eq. (10.45) can then be written as,

$$K_\delta\left(x,t \mid x_0,0\right) = K_0\left(x,t \mid x_0,0\right)$$

$$- \frac{\gamma}{\sqrt{2\pi}} e^{(i/2t)(|x|+|x_0|)^2} \int_0^{\infty} e^{-\left(z+\gamma\sqrt{it}\right)y-\left(y^2/2\right)} dy$$

$$= K_0\left(x,t \mid x_0,0\right)$$

$$- \frac{\gamma}{2} e^{(i/2t)(|x|+|x_0|)^2} e^{\left[\frac{1}{2}\left(z+\gamma\sqrt{it}\right)^2\right]} \left[1 - \phi\left(\frac{z}{\sqrt{2}} + \gamma\sqrt{\frac{it}{2}}\right)\right],$$

$$\quad (10.48)$$

where the probability integral is given from 8.252(6) of [98], for Re $\eta^2 > 0$,

$$\phi(y/2\eta) = 1 - (2/\sqrt{\pi})\eta e^{-y^2/4\eta^2} \int_0^{\infty} e^{-t^2\eta^2 - ty} dt. \quad (10.49)$$

In terms of the complementary error function,

$$Erfc\left(z\right) = 1 - Erf\left(z\right) = 1 - \phi(z) \quad (10.50)$$

with

$$Erf(z) = (2/\sqrt{\pi}) \int_0^z \exp\left(-s^2\right) ds, \qquad (10.51)$$

we obtain the propagator for a delta function potential [16],

$$K_\delta\left(x,t \mid x_0, 0\right) = K_0\left(x,t \mid x_0, 0\right)$$

$$- \frac{\gamma}{2} e^{\gamma(|x|+|x_0|)+(i\gamma^2 t/2)} Erfc\left[\left(\frac{|x|+|x_0|}{\sqrt{2it}}\right) + \gamma\sqrt{\frac{it}{2}}\right]. \qquad (10.52)$$

Using Eq. (10.52) in the linear combination of image propagators in Eq. (10.40), we obtain the following result for the free particle propagator with the general boundary condition, $\left[(\partial/\partial x) K_\delta^{\beta/\alpha}\left(x,t \mid x_0, 0\right)\right]_{x=0} = \gamma K_\delta^{\beta/\alpha}\left(0,t \mid x_0, 0\right)$, as the expectation $TI_\delta(0) \equiv K_\delta^{\beta/\alpha}(x,t \mid x_0, 0)$,

$$K_\delta^{\beta/\alpha}\left(x,t \mid x_0, 0\right) = \frac{1}{\sqrt{2\pi it}}\left(e^{\frac{i}{2t}(x-x_0)^2} + e^{\frac{i}{2t}(x+x_0)^2}\right)$$

$$- \gamma e^{\gamma(|x|+|x_0|)+(i\gamma^2 t/2)} Erfc\left[\left(\frac{|x|+|x_0|}{\sqrt{2it}}\right) + \gamma\sqrt{\frac{it}{2}}\right]. \qquad (10.53)$$

As a linear combination of $K_\delta\left(x,t \mid x_0, t_0\right)$ and $K_\delta\left(-x,t \mid x_0, t_0\right)$, each of which satisfies the Schrödinger equation, $K_\delta^{\beta/\alpha}\left(x,t \mid x_0, 0\right)$ then solves the Schrödinger equation,

$$\left[i\partial_t + \frac{1}{2}(\partial/\partial x)^2\right] K_\delta^{\beta/\alpha}\left(x,t \mid x_0, 0\right) = 0; \quad x > 0. \qquad (10.54)$$

The full propagator of Eq. (10.53) matches the result of Clark, Menikoff and Sharp [58]. In the limit $t \to 0$,

$$Erfc\left[\left(\frac{|x|+|x_0|}{\sqrt{2it}}\right) + \gamma\sqrt{\frac{it}{2}}\right] = 0, \qquad (10.55)$$

and the first two terms in Eq. (10.53) become $\delta\left(x - x_0\right)$ and $\delta\left(x + x_0\right)$, respectively. Since the physical problem has endpoints x and x_0 located in the half-line $x \geq 0$, then $\delta\left(x + x_0\right)$ is simply zero, and the propagator satisfies the initial condition in the region of interest,

$$\lim_{t \to 0} K_\delta^{\beta/\alpha}\left(x,t \mid x_0, 0\right) = \delta\left(x - x_0\right). \qquad (10.56)$$

For the special case of Dirichlet boundary conditions, Eq. (10.52) reduces to the propagator for delta interaction models in earlier works. See, for example, the comprehensive study and review of Schrödinger operators done by S. Albeverio *et al.* [5], as well as Feynman path integral treatments cited in the Handbook of Feynman Integrals of Grosche and Steiner [99].

Exercises

(10-1) Using the path integral method, calculate the popagator of a particle constrained to move on the surface of an infinite cylinder of radius R_0.

(10-2) Show explicitly that the propagator, Eq. (10.53), satisfies Eq. (10.56).

Chapter 11

Relativistic Quantum Mechanics

An electron moving at half the speed of light requires a description which is not only quantum mechanical, but also relativistic. Nature is replete with systems appropriately described by a quantum relativistic theory. The fusion of quantum mechanics and special relativity is elegantly exemplified by the Dirac equation for spin $\frac{1}{2}$ particles, and the Klein-Gordon equation for spin 0 particles. In general, to understand quantum relativistic systems, one aims to find solutions to these equations especially for particles in an external field. One way of obtaining solutions is the path integral method of Feynman which, in non-relativistic quantum mechanics, has been a powerful alternative to Schrödinger's wave mechanics and Heisenberg's matrix mechanics. There are, in fact, various procedures in applying the Feynman path integral to solve the quantum propagator of relativistic particles. In this chapter, we choose to adopt the proper time formulation earlier discussed by V. Fock [87], R. P. Feynman [83; 84; 85], Y. Nambu [161], C. Morette [159], J. Schwinger [186], and later by B. S. DeWitt [66; 67], among others. An advantage of this formulation is that one can cast the path integral into a form quite similar to those which appear in non-relativistic path integrals. The techniques developed for the nonrelativistic case, therefore, can be utilized, and an exact path integral evaluation is possible which yields explicit forms for the propagator of various quantum relativistic systems.

Section 11.1 extends the applicability of the Hida-Streit white noise path integral to solve systems in relativistic quantum mechanics where the Green function for the Dirac equation can be obtained. We then take as an example in Section 11.2 the case of a relativistic charged particle in a uniform magnetic field. The paths of the particle are parametrized in terms of the white noise variable ω and the effective potential becomes similar in form to that of Lévy's stochastic area. As in the nonrelativistic case, the path integral can be evaluated with the help of the T-transform in white noise analysis from which we obtain the Dirac Green function.

11.1   Green Function for the Dirac Equation

In 1948, when R. P. Feynman introduced path integration as an alternative way of solving problems in non-relativistic quantum mechanics, the last section of his pioneering paper also contained a discussion on how to treat the Klein-Gordon equation and particles with spin [83]. Early attempts to apply the path integral method to solve quantum systems in special and general relativity theories followed [84; 85; 157; 159], but further developments had to wait until methods and techniques, most especially for non-Gaussian path integrals, were developed in the non-relativistic case.

Consider the Dirac equation for a spin $\frac{1}{2}$ particle of mass m given by,

$$\left(m - \widehat{M} \right) G\left(\mathbf{r}'', \mathbf{r}' \right) = \delta\left(\mathbf{r}'' - \mathbf{r}' \right), \tag{11.1}$$

where we have defined an operator $\widehat{M}$ in terms of the Dirac matrices $\boldsymbol{\alpha}$ and β of the form,

$$\widehat{M} = -\beta\boldsymbol{\alpha}\cdot\left(\mathbf{p} - e\mathbf{A} \right) - \beta V + \beta E. \tag{11.2}$$

Here, V and $\mathbf{A}$ are the scalar and vector potentials, respectively. By expressing the Green function $G\left(\mathbf{r}'', \mathbf{r}' \right)$ in Eq. (11.1) as,

$$G\left(\mathbf{r}'', \mathbf{r}' \right) = \left(m + \widehat{M} \right) g\left(\mathbf{r}'', \mathbf{r}' \right), \tag{11.3}$$

we obtain an iterated Dirac equation,

$$\left(m^2 - \widehat{M}^2 \right) g\left(\mathbf{r}'', \mathbf{r}' \right) = \delta\left(\mathbf{r}'' - \mathbf{r}' \right). \tag{11.4}$$

Note that the solutions of Eq. (11.1) are also solutions of Eq. (11.4), but not conversely.

The Green functions, $G\left(\mathbf{r}'', \mathbf{r}' \right) = \langle \mathbf{r}''|G|\mathbf{r}' \rangle$ and $g\left(\mathbf{r}'', \mathbf{r}' \right) = \langle \mathbf{r}''|g|\mathbf{r}' \rangle$, are matrix elements of the operators, $G = (m - \widehat{M})^{-1}$, and $g = (m^2 - \widehat{M}^2)^{-1}$, respectively. We can also write, for example operator g, in integral form using the relation (see, e.g., [83], Eq. (5-17)),

$$\lim_{\epsilon \to 0} \left[-(\theta + i\epsilon)^{-1} \right] = i\int\limits_0^\infty \exp\left(i\theta\Lambda \right) d\Lambda. \tag{11.5}$$

Defining

$$H = \left(m^2 - \widehat{M^2}\right)/2m\,, \tag{11.6}$$

we can express the operator $g = (1/2m)\,(1/H)$ as (see, also, [65; 186])

$$g = (i/2m) \int_0^\infty \exp\left(-iH\Lambda\right) d\Lambda\,. \tag{11.7}$$

Taking the matrix element of Eq. (11.7),

$$g\left(\mathbf{r}'',\mathbf{r}'\right) = (i/2m) \int_0^\infty \langle \mathbf{r}''|\exp\left(-iH\Lambda\right)|\mathbf{r}'\rangle\, d\Lambda\,, \tag{11.8}$$

we observe that the integrand in Eq. (11.8) is analogous in form to a quantum propagator evolving in Λ-time with an effective Hamiltonian H. Following Feynman's prescription for handling quantum propagators [83] we can express the integrand $\langle \mathbf{r}''|\exp\left(-iH\Lambda\right)|\mathbf{r}'\rangle$ as a path integral (see Eqs. (9.8) and (9.11)). After evaluating the propagator $\langle \mathbf{r}''|\exp\left(-iH\Lambda\right)|\mathbf{r}'\rangle$, we can integrate out Λ as in Eq. (11.8) to get $g\left(\mathbf{r}'',\mathbf{r}'\right)$, and from Eq. (11.3), the Green function $G\left(\mathbf{r}'',\mathbf{r}'\right)$ for the Dirac equation can then be subsequently calculated. This procedure, therefore, enables us to arrive at a solution for the Dirac equation via the path integral method. We evaluate as an example of using white noise analytical methods the case of a charged particle in a uniform magnetic field.

11.2 Dirac Particle in a Uniform Magnetic Field

Let us now consider an electron of charge e ($e < 0$) and mass m subjected to a uniform magnetic field $\mathbb{B}$ along the z-axis and represented by the vector potential, $\mathbf{A} = (1/2)\,(-\mathcal{B}y, \mathcal{B}x, 0)$. Here, we represent the magnetic field by, $\mathbb{B} = \mathcal{B}\hat{k}$, to distinguish it from our notation for the Brownian motion, B. With $\widehat{M} = -\beta\boldsymbol{\alpha}\cdot(\mathbf{p} - e\mathbf{A}) + \beta E$ from Eq. (11.2), the operator H, Eq. (11.6), has the form,

$$H = (1/2m)\left[(p_x + e\mathcal{B}y/2)^2 + (p_y - e\mathcal{B}x/2)^2 + p_z^2\right]$$
$$- \left(k_0^2/2m\right) - (e\boldsymbol{\sigma}\cdot\mathbb{B}/2m)\,, \tag{11.9}$$

where $k_0^2 = E^2 - m^2$. The spin part, with the magnetic field $\mathbb{B}$ along the z-axis, is just $-e\boldsymbol{\sigma}\cdot\mathbb{B}/2m = -e\mathcal{B}\sigma_z/2m$. We can then use Eq. (11.9) as the effective Hamiltonian in $\langle \mathbf{r}''|\exp\left(-iH\Lambda\right)|\mathbf{r}'\rangle$ of Eq. (11.8). Defining the operator $S_z = (1/2)\,\sigma_z$, then S_z acting upon an eigenspinor has the eigenvalues, $s_i = \pm 1/2$. Noting that $|\mathbf{r}\rangle = |xyz\rangle\,|\chi\rangle$, with $|\chi\rangle$ a spinor, we can use $\sum |s_i\rangle\,\langle s_i| = 1$, to write the integrand in Eq. (11.8) as,

$$\langle \mathbf{r}''| \exp\left(-iH\Lambda\right) |\mathbf{r}'\rangle \equiv \sum_{s_i} \langle \chi|s_i\rangle\,\langle x''y''z''| \exp\left(-iH_s\Lambda\right) |x'y'z'\rangle\,\langle s_i|\chi\rangle$$

$$= \sum_{s_i} \eta_{s_i}\eta_{s_i}^+ \langle x''y''z''| \exp\left(-iH_s\Lambda\right) |x'y'z'\rangle \qquad (11.10)$$

where $\eta_{s_i} = \langle \chi|s_i\rangle$. The coordinate part, $\langle x''y''z''| \exp\left(-iH_s\Lambda\right) |x'y'z'\rangle$, has an effective Hamiltonian of the form (setting $\gamma = e\mathcal{B}/2$),

$$H_s = (1/2m)\left[(p_x + e\mathcal{B}y/2)^2 + (p_y - e\mathcal{B}x/2)^2 + p_z^2\right]$$

$$- \left(k_0^2/2m\right) - \left(2\gamma s_i/m\right). \qquad (11.11)$$

We then evaluate the coordinate part as the path integral

$$\langle x''y''z''| \exp\left(-iH_s\Lambda\right) |x'y'z'\rangle = \int \exp\left(iS\right) D\left[xyz\right], \qquad (11.12)$$

for a system evolving with a time-like parameter Λ, and an action corresponding to Eq. (11.11) given by,

$$S = \int\limits_0^\Lambda \left[\frac{1}{2}m\left(\dot{x}^2 + \dot{y}^2 + \dot{z}^2\right) + \gamma\left(x\dot{y} - y\dot{x}\right) + \left(k_0^2/2m\right) + \left(2\gamma s_i/m\right)\right] d\lambda.$$

$$(11.13)$$

The path integral, Eq. (11.12), can be written as,

$$\langle x''y''z''| \exp\left(-iH_s\Lambda\right) |x'y'z'\rangle = e^{i\left[\left(k_0^2/2m\right)+\left(2\gamma s_i/m\right)\right]\Lambda}$$

$$\times K\left(x'', y''; x', y'\right) K\left(z'', z'\right).$$

$$(11.14)$$

In Eq. (11.14), the propagator along the z-coordinate is similar to the free particle case, i.e.,

$$K\left(z'', z'\right) = \int \exp\left[(im/2)\,\dot{z}^2\right] D\left[z\right]$$

$$= (1/2\pi) \int \exp\left\{\left[ik_z\left(z'' - z'\right) - \left(ik_z^2/2m\right)\right]\Lambda\right\} dk_z.$$

$$(11.15)$$

On the other hand, the propagator along the (x, y) plane is,

$$K\left(x''y''; x'y'\right) = \int \exp\left\{i\int_0^\Lambda \left[\frac{1}{2}m\left(\dot{x}^2 + \dot{y}^2\right) + \gamma\left(x\dot{y} - y\dot{x}\right)\right] d\lambda\right\} D\left[xy\right].$$

(11.16)

This equation is similar to the propagator for a nonrelativistic charged particle in a uniform magnetic field [61; 83] which can be evaluated using white noise calculus.

The evaluation of $K\left(x'', y''; x', y'\right)$ has been done previously for the nonrelativistic case (see Eq. (9.82)). In terms of the Hermite polynomials H_n, we obtain,

$$K\left(x'', y''; x', y'\right) = (\gamma/\pi) \exp\left\{i\gamma\left(x'y'' - x''y'\right)\right\}$$

$$\times \exp\left\{(-\gamma/2)\left[(x'' - x')^2 + (y'' - y')^2\right]\right\}$$

$$\times \sum_{n=0}^{\infty} \sum_{q=0}^{n} \frac{(-1)^{-n} 2^{-2n}}{q!\,(n-q)!}$$

$$\times H_{2q}\left(\sqrt{\gamma}\left(x'' - x'\right)\right) H_{2(n-q)}\left(\sqrt{\gamma}\left(y'' - y\right)\right)$$

$$\times \exp\left[-i\left(n + \frac{1}{2}\right) 2\gamma\Lambda\right].$$

(11.17)

Using Eqs. (11.15) and (11.17), we can now write Eq. (11.14) as,

$$\langle x''y''z''|\exp\left(-iH_s\Lambda\right)|x'y'z'\rangle$$

$$= \left(\gamma/2\pi^2\right) \exp\left\{i\gamma\left(x'y'' - x''y'\right)\right\}$$

$$\times \exp\left\{(-\gamma/2)\left[(x'' - x')^2 + (y'' - y')^2\right]\right\}$$

$$\times \int dk_z \exp\left[ik_z\left(z'' - z'\right)\right] \sum_{n=0}^{\infty} \sum_{q=0}^{n} [(-1)^{-n} 2^{-2n}/q!$$

$$\times (n-q)!] H_{2q}\left[\sqrt{\gamma}\left(x'' - x'\right)\right] H_{2(n-q)}$$

$$\times \left[\sqrt{\gamma}\left(y'' - y'\right)\right] \exp\left[-i\left(k_z^2 - b^2\right)\Lambda/2m\right],$$

(11.18)

where $b^2/2m = \left(k_0^2/2m\right) + \left[s_i - \left(n + \frac{1}{2}\right)\right] (2\gamma/m)$ (with the mass m). From Eqs. (11.10) and (11.18), the Green function for the iterated Dirac equation

(11.8), appears as,

$$g\left(\mathbf{r}'',\mathbf{r}'\right) = \left[i\gamma/m\left(2\pi\right)^2 \int dk_z \sum_{s_i} \eta_{s_i}\eta_{s_i}^+ \exp\left[ikz\left(z''-z'\right)\right]\right]$$

$$\times \exp\left\{i\gamma\left(x'y''-x''y'\right)\right\}\exp\left\{(-\gamma/2)\right.$$

$$\times \left[\left(x''-x'\right)^2 + \left(y''-y'\right)^2\right]\right\}\sum_{n=0}^{\infty}\sum_{q=0}^{n}\left[\frac{(-1)^{-n}\,2^{-2n}}{q!\,(n-q)!}\right]$$

$$\times H_{2q}\left[\sqrt{\gamma}\left(x''-x'\right)\right]H_{2(n-q)}\left[\sqrt{\gamma}\left(y''-y'\right)\right]$$

$$\times \int_0^{\infty} \exp\left[-i\left(k_z^2-b^2\right)\Lambda/2m\right]d\Lambda. \qquad (11.19)$$

The integration over Λ yields the result [98],

$$g\left(\mathbf{r}'',\mathbf{r}'\right) = \left[i\gamma/m\left(2\pi\right)^2\right] \int dk_z \sum_{s_i} \eta_{s_i}\eta_{s_i}^+ \exp\left[ikz\left(z''-z'\right)\right]$$

$$\times \exp\left\{i\gamma\left(x'y''-x''y'\right)\right\}\exp\left\{(-\gamma/2)\right.$$

$$\times \left[\left(x''-x'\right)^2 + \left(y''-y'\right)^2\right]\right\}\sum_{n=0}^{\infty}\sum_{q=0}^{n}\left[\frac{(-1)^{-n}\,2^{-2n}}{q!\,(n-q)!}\right]$$

$$\times H_{2q}\left[\sqrt{\gamma}\left(x''-x'\right)\right]H_{2(n-q)}\left[\sqrt{\gamma}\left(y''-y'\right)\right]$$

$$\times \left\{(1/2m)\left(E^2-m^2-k_z^2\right) + \left[s_i-\left(n+\frac{1}{2}\right)\right](2\gamma/m)+i\varepsilon\right\}^{-1},$$

$$(11.20)$$

where $\varepsilon \to 0$. At this point, the discrete energy spectrum [113] can already be obtained from the poles of the Green function and is given by, $E^2 = m^2 + k_z^2 + [(2n+1)-2s_i]\,e\mathcal{B}$.

Finally, the Green function $G\left(\mathbf{r}'',\mathbf{r}'\right)$ for the first-order Dirac equation can be obtained with the help of Eqs. (11.3) and (11.20). From Eq. (11.2), we can write $\widehat{M}$ (with $V=0$) as,

$$\widehat{M} = -\beta\alpha_x\left(-i\partial/\partial x'' + e\mathcal{B}y''/2\right) - \beta\alpha_y\left(-i\partial/\partial y'' - e\mathcal{B}x''/2\right)$$

$$+ \left(i\beta\alpha_z\partial/\partial z''\right) + \beta E \qquad (11.21)$$

and using the recursion relation of the Hermite polynomials,

$$\frac{dH_n(x)}{dx} = 2nH_{n-1}(x) \tag{11.22}$$

we obtain,

$$G(\mathbf{r}'', \mathbf{r}') = \left\{ m - \left(\frac{eB}{2}\right) \beta \alpha_+ [(y'' - y') + i(x'' - x') + \beta E - k\beta \alpha_z \right\}$$

$$\times g(\mathbf{r}'', \mathbf{r}') + i\beta \alpha_x \bar{g} \sqrt{\frac{eB}{2}} 4q$$

$$\times H_{2q-1}\left[\sqrt{\frac{eB}{2}}(x'' - x')\right] H_{2(n-q)}\left[\sqrt{\frac{eB}{2}}(y'' - y')\right]$$

$$+ i\beta \alpha_y \bar{g} \sqrt{\frac{eB}{2}} 4(n-q)$$

$$\times H_{2(n-q)-1}\left[\sqrt{\frac{eB}{2}}(y'' - y')\right] H_{2q}\left[\sqrt{\frac{eB}{2}}(x'' - x')\right], \tag{11.23}$$

where

$$\bar{g} = \left[ieB/2m(2\pi)^2\right] \int dk_z \sum_{s_i} \eta_{s_i} \eta_{s_i}^+ \exp\left[z'' - z'\right]$$

$$\times \exp\left\{(ieB/2)(x'y'' - x''y') - (eB/4)\left[(x'' - x')^2 + (y'' - y')^2\right]\right\}$$

$$\times \sum_{n=0}^{\infty}\sum_{q=0}^{n}\left[(-1)^{-n} 2^{-2n}/q!(n-q)!\right]$$

$$\times \left\{(1/2m)(E^2 - m^2 - k_z^2) + \left[s_i - \left(n + \frac{1}{2}\right)\right](eB/m) + i\varepsilon\right\}^{-1} \tag{11.24}$$

with $\varepsilon \to 0$. We note here that an early attempt to provide a path integral treatment of a Dirac particle in a uniform magnetic field can be found in [167]. In general, the path integral method has been used to solve the Klein-Gordon equation (see, e.g., [18; 19]) and the Dirac equation (see, e.g., [32; 118]). Hence, further utilization of white noise calculus for relativistic problems can be more fully explored [20].

Exercises

(11-1) Consider the interated Dirac equation for a particle of mass μ,

$$\left(\mu^2 - \widehat{M^2}\right) |\Phi\rangle = 0 \,,$$

where $\widehat{M} = i\gamma^\mu \left(\partial_\mu + ieA_\mu\right)$, $\gamma^\mu = (\beta, \beta\overline{\alpha})$, $\partial_\mu = (\partial/\partial t, \nabla)$ and $A_\mu = (V, -\mathbf{A})$. Derive the expression,

$$\widehat{M^2} = \left(i\partial_\mu - eA_\mu\right)^2 - (e/2)\,\sigma^{\mu v} F_{\mu v} \,,$$

where $\sigma_{\mu v} = (i/2)\left(\gamma_\mu \gamma_v - \gamma_v \gamma_\mu\right)$ and $F_{\mu v} = \partial_\mu A_v - \partial_v A_\mu$.

(11-2) Show that the spin-dependent term in **(11-1)** may be written as,

$$(e/4)\,\sigma^{\mu v} F_{\mu v} = (e/2)\left(i\boldsymbol{\alpha} \cdot \mathbf{E} + \boldsymbol{\sigma} \cdot \mathbf{B}\right),$$

where $\mathbf{E}$ and $\mathbf{B}$ are the electric and magnetic fields, respectively.

Appendix A

Useful Integrals

A.1   Integrals with Gaussian White Noise Measure

We start with the characteristic functional:

$$\int_{S_*} \exp\left(i \langle \omega, \xi \rangle\right) \, d\mu(\omega) = \exp\left(-\frac{1}{2}\int \xi^2 d\tau\right), \qquad \xi \in S.\qquad \text{(A.1)}$$

An important integral involves the Gauss kernel:

$$\int_{S_*} \mathcal{N} \, \exp\left(i \langle \omega, \xi \rangle\right) \exp\left(-\frac{1}{2}\langle \omega, k\omega \rangle\right) \, d\mu(\omega) = \exp\left[-\frac{1}{2}\langle \xi, (k+1)^{-1}\xi \rangle\right],$$

$$\text{(A.2)}$$

where $\mathcal{N}$ in Eq. (A.2) is given by,

$$\mathcal{N}^{-1} = \int \exp\left[-\frac{k}{2}\int \omega^2 d\tau\right] d\mu(\omega).\qquad \text{(A.3)}$$

From Eq. (A.2), we have a special case $k = -(i+1)$, useful in quantum mechanics:

$$\int_{S_*} \mathcal{N} \exp\left(i \langle \omega, \xi \rangle\right) \exp\left[\left(\frac{i+1}{2}\right)\int \omega(\tau)^2 \, d\tau\right] \, d\mu(\omega) = \exp\left(-\frac{i}{2}\int \xi^2 d\tau\right),$$

$$\text{(A.4)}$$

where $\mathcal{N}$ in Eq. (A.4) is given by,

$$\mathcal{N}^{-1} = \int \exp\left[\frac{(i+1)}{2}\int \omega^2 d\tau\right] d\mu(\omega).\qquad \text{(A.5)}$$

From Eq. (2.13) of [196]:

$$\int_{S^*} \exp\left(i\left\langle\omega,\xi\right\rangle\right)\ \exp\left(-\frac{1}{2}\left\langle\omega,K\omega\right\rangle\right)\exp\left(-\frac{1}{2}\left\langle\omega,L\omega\right\rangle\right)\ d\mu(\omega)$$

$$= \exp\left[-\frac{1}{2}Tr\ln\left(1 + L\left(1+K\right)^{-1}\right)\right]$$

$$\times \exp\left[-\frac{1}{2}\left\langle\xi,\left(1+K+L\right)^{-1}\xi\right\rangle\right]. \tag{A.6}$$

For $K = 0$ and $\xi = 0$ in Eq. (A.6), we also have,

$$\int \exp\left(-\frac{1}{2}\left\langle\omega,L\ \omega\right\rangle\right)\ d\mu\left(\omega\right) = \left[\det\left(1+L\right)\right]^{-1/2}. \tag{A.7}$$

An integral needed for Lévy's stochastic area is:

$$\int \exp\left[izS_T\right]\ d\mu\left(\omega\right) = \left\{\prod_{n=1}^{\infty}\left[1 + \frac{4z^2T^2}{4\left(2n-1\right)^2\pi^2}\right]^2\right\}^{-1/2}$$

$$= \left[\cosh\left(zT/2\right)\right]^{-1}, \tag{A.8}$$

with S_T given by [104; 142],

$$S_T = \frac{1}{2}\int_0^T \left[B_x\left(t\right)\ dB_y\left(t\right) - B_y\left(t\right)\ dB_x\left(t\right)\right], \tag{A.9}$$

where $\mathbf{B}(t) = \left(B_x\left(t\right),B_y\left(t\right)\right)$ is a two-dimensional Brownian motion in the time interval $0 \le t \le T$. Equation (A.8) appears in problems with uniform magnetic field.

Bibliography

[1] L. F. Abbott, E. Farhi, and S. Gutmann, The path integral for dendritic trees, *Biol. Cyber.* **66** (1991) 49-60.

[2] A. Aldo Faisal, L. P. J. Selen, and D. M. Wolpert, Noise in the nervous system, *Nat. Rev. Neurosci.* **9** (2008) 292-303.

[3] M. Abramowitz and I. A. Stegun, *Handbook of Mathematical Functions with Formulas, Graphs, and Mathematical Tables* (National Bureau of Standards, Washington D. C., 1972).

[4] Y. Aharonov and D. Bohm, Significance of electromagnetic potentials in quantum theory, *Phys. Rev.* **115** (1959) 485-491.

[5] S. Albeverio, F. Gesztesy, R. Høegh-Krohn, and H. Holden, *Solvable Models in Quantum Mechanics* (Springer-Verlag, Berlin, 1988).

[6] S. Albeverio, F. Gesztesy, R. Høegh-Krohn, and L. Streit, Charged particles with short range interactions, *Ann. Inst. Henri Poincare* A **38** (1983) 263-293.

[7] S. Albeverio, R. Høegh-Krohn, and S. Mazzucchi, *Mathematical Theory of Feynman Path Integrals - An Introduction*, 2nd ed. (Springer-Verlag, Berlin, 2008).

[8] S. Alfarano and M. Milaković, Does classical competition explain the statistical features of firm growth?, *Econ. Lett.* **101** (2008) 272-274.

[9] J. Alvarez-Ramirez, J. Alvarez, E. Rodriguez, and G. Fernandez-Anaya, Time-varying Hurst exponent for US stock markets, *Physica A: Stat. Mech. Its Appl.* **387** (2008) 6159-6169.

[10] L. A. N. Amaral, P. Ch. Ivanov, N. Aoyagi, I. Hidaka, S. Tomono, A. L. Goldberger, H. E. Stanley, and Y. Yamamoto, Behavioral-independent features of complex heartbeat dynamics, *Phys. Rev. Lett.* **86** (2001) 6026-6029.

[11] H. P. Aringa, C. C. Bernido, M. V. Carpio-Bernido, and J. B. Bornales, Stochastic modelling of helical biopolymers, *Int. J. Mod. Phys. Conf. Ser.* **17** (2012) 73-76.

[12] L. Bachelier, Théorie de la speculation, *Annales Scientifiques de l'École Normale Supérieure 3*, **17** (1900) 21-86.

[13] J. Balakrishnan, Neural network learning dynamics in a path integral framework, arXiv:cond-mat/0308503v1 [cond-mat.stat-mech] (2003).

[14] M. Bartos, I. Vida, and P. Jonas, Synaptic mechanisms of synchronized gamma oscillations in inhibitory interneuron networks, *Nature Rev. Neurosci.* **8** (2007) 45-56.

[15] R. A. D. Bathgate, C. S. Samuel, T. C. D. Burazin, A. L. Gundlach, and G. W. Tregear, Relaxin: new peptides, receptors, and novel actions, *Trends in Endocrinology and Metabolism* **14** (2003) 207-213.

[16] D. Bauch, The path integral for a particle moving in a δ-function potential, *Nuovo Cimento B* **85** (1985) 118-124.

[17] R. Bellman, *A Brief Introduction to Theta Functions* (Holt, Rinehart & Winston, New York, 1961).

[18] C. C. Bernido, Path integral treatment of the gravitational anyon in a uniform magnetic field, *J. Phys. A: Math. Gen.* **26** (1993) 5461-5471.

[19] C. C. Bernido and G. Aguarte, Summation over histories for a particle in spherical orbit around a black hole, *Phys. Rev. D* **56** (1997) 2445-2448.

[20] C. C. Bernido, J. B. Bornales and M. V. Carpio-Bernido, Application of white noise calculus in evaluating the path integral in relativistic quantum mechanics, *Comm. Stoch. Anal.* **1** (2007) 151-161.

[21] C. C. Bernido and M. V. Carpio-Bernido, Path integrals for boundaries and topological constraints: A white noise functional approach, *J. Math. Phys.* **43** (2002) 1728-1736.

[22] C. C. Bernido and M. V. Carpio-Bernido, Entanglement probabilities of polymers: a white noise functional approach, *J. Phys A: Math. Gen.* **36** (2003) 4247-4257.

[23] C. C. Bernido and M. V. Carpio-Bernido, Overwinding in a stochastic model of an extended polymer, *Phys. Lett. A* **369** (2007) 1-4.

[24] C. C. Bernido and M. V. Carpio-Bernido, On a fractional stochastic path integral approach in modelling interneuronal connectivity, *Int. J. Mod. Phys. Conf. Ser.* **17** (2012) 23-33.

[25] C. C. Bernido and M. V. Carpio-Bernido, White noise analysis: some applications in complex systems, biophysics and quantum mechanics, *Int. J. Mod. Phys. B* **26** (2012) 1230014.

[26] C. C. Bernido and M. V. Carpio-Bernido, Transition probabilities for processes with memory on topological non-trivial spaces, in *Stochastic and Infinite Dimensional Analysis*, eds. C. C. Bernido, M. V. Carpio-Bernido, M. Grothaus, T. Kuna, J. L. da Silva, and M. J. Oliveira (Birkhäuser, Basel, 2014).

[27] C. C. Bernido, M. V. Carpio-Bernido, and H. P. Aringa, Topology-dependent entropic differences for chiral polypeptides in solvents, *Phys. Lett. A* **375** (2011) 1225-1228.

[28] C. C. Bernido, M. V. Carpio-Bernido, and J. B. Bornales, On chirality and length-dependent potentials in polymer entanglements, *Phys. Lett. A* **339** (2005) 232-236.

[29] C. C. Bernido, M. V. Carpio-Bernido, and M. G. O. Escobido, Modeling protein conformations: Wiegel's helical polymer and the Hida-Streit white

noise path integral, in *Stochastic and Quantum Dynamics of Biomolecular Systems*, eds. C. C. Bernido and M. V. Carpio-Bernido (American Inst. of Physics, New York, 2008), pp. 139-148.

[30] C. C. Bernido, M. V. Carpio-Bernido, and M. G. O. Escobido, Modified diffusion with memory for cyclone track fluctuations, *Phys. Lett. A* **378** (2014) 2016-2019.

[31] C. C. Bernido, M. V. Carpio-Bernido, and A. Inomata, On evaluating topologically constrained path integrals *Phys. Lett. A* **136** (1989) 259.

[32] C. C. Bernido, M. V. Carpio-Bernido and N. S. Lim, Path integral quantization of a Dirac–Coulomb problem on the half-line with superstrong magnetic fields, *Phys. Lett. A* **231** (1997) 395-401.

[33] C. C. Bernido, M. V. Carpio-Bernido, and L. Streit, in preparation.

[34] C. C. Bernido, M. G. O. Escobido, and M. V. Carpio-Bernido, Fractional path integration for growth dynamics and scaling behavior in complex systems, *Int. J. Mod. Phys. Conf. Ser.* **17** (2012) 94-103.

[35] C. C. Bernido and A. Inomata, Topological shifts in the Aharonov-Bohm effect, *Phys. Lett. A* **77** (1980) 394.

[36] C. C. Bernido and A. Inomata, Path integrals with a periodic constraint: The Aharonov-Bohm effect, *J. Math. Phys.* **22** (1981) 715.

[37] F. Biagini, Y. Hu, B. Øksendal, and T. Zhang, *Stochastic Calculus for Fractional Brownian Motion and Applications* (Springer-Verlag, London, 2008).

[38] T. Binzegger, R. J. Douglas, and K. A. Martin, A quantitative map of the circuit of cat primary visual cortex, *J. Neurosci.* **24** (2004) 8441-8453.

[39] P. Blanchard, Ph. Combe, H. Nencka, and R. Rodriguez, Stochastic dynamical aspects of neuronal activity, *J. Math. Biol.* **31** (1993) 189-198.

[40] P. Bonifazi, M. Goldin, M. A. Picardo, I. Jorquera, A. Cattani, G. Bianconi, A. Represa, Y. Ben-Ari, and R. Cossart, GABAergic hub neurons orchestrate synchrony in developing hippocampal networks, *Science* **326** (2009) 1419.

[41] G. Bottazzi and A. Secchi, Why are distributions of firm growth rates tent-shaped?, *Econ. Lett.* **80** (2003) 415-420.

[42] V. Braitenberg and A. Schüz, *Statistics and Geometry of Neuronal Connectivity* (Springer-Verlag, Berlin, 1998).

[43] P. J. Brockwell and R. A. Davis, *Introduction to Time Series and Forecasting*, 2nd ed. (Springer-Verlag, New York, 2002).

[44] R. M. Bryce and K. B. Sprague, Revisiting detrended fluctuation analysis, *Sci. Rep.* **2** (2012) 315; DOI:10.1038/srep00315 (2012).

[45] E. Bullmore, C. Long, J. Suckling, J. Fadili, G. Calvert, F. Zelaya, T. A. Carpenter, and M. Brammer, Colored noise and computational inference in neurophysiological (fMRI) time series analysis: resampling methods in time and wavelet domains, *Human Brain Mapping* **12** (2001) 61-78.

[46] I. Calvo and R. Sánchez, The path integral formulation of fractional Brownian motion for the general Hurst exponent, *J. Phys. A: Math. Theor.* **41** (2008) 282002.

[47] S. J. Camargo, A. W. Robertson, S. J. Gaffney, P. Smyth, and M. Ghil,

Cluster analysis of typhoon tracks. Part I: General properties, *J. Climate* **20** (2007) 3635-3653.

[48] A. Carbone, G. Castelli, and H. E. Stanley, Time-dependent hurst exponent in financial time series, *Physica A: Stat. Mech. Its Appl.* **344** (2004) 267-271.

[49] M. V. Carpio-Bernido, Path integral quantization of certain noncentral systems with dynamical symmetries, *J. Math. Phys.* **32** (1991) 1799-1807.

[50] M. V. Carpio-Bernido, Green's function for an axially symmetric potential: an evaluation in polar coordinates, *J. Phys. A: Math. Gen.* **24** (1991) 3013-3019.

[51] M. V. Carpio-Bernido and C. C. Bernido, An exact solution for a ring-shaped oscillator plus a $c \sec^2 \vartheta / r^2$ potential, *Phys. Lett. A* **134** (1989) 395-399.

[52] M. V. Carpio-Bernido and C. C. Bernido, Algebraic treatment of the double ring-shaped oscillator, *Phys. Lett. A* **137** (1989) 1-3.

[53] B. Cessac and M. Samuelides, From neuron to neural networks dynamics, *EPJ Special Topics* **142** (2007) 5-88.

[54] J. C. L. Chan and J. D. Kepert, eds. *Global Perspectives on Tropical Cyclones* (World Scientific, Singapore, 2010).

[55] S. Chandrasekhar, Stochastic problems in physics and astronomy, *Rev. Mod. Phys.* **15** (1943) 1-89.

[56] E. J. Chichilnisky, A simple white noise analysis of neuronal light responses, *Network: Computation in Neural Systems* **12** (2001) 199-213.

[57] Y. H. Ch'ng and R. C. Reid, Cellular imaging of visual cortex reveals the spatial and functional organization of spontaneous activity, *Front. Integr. Neurosci.* **4** (2010) 1-9.

[58] T. E. Clark, R. Menikoff, and D. H. Sharp, Quantum mechanics on the half-line using path integrals, *Phys. Rev. D* **22** (1980) 3012.

[59] H. Cui and R. A. Andersen, Posterior parietal cortex encodes autonomously selected motor plans, *Neuron* **56** (2007) 552.

[60] M. Cunha, C. Drumond, P. Leukert, J. L. da Silva, and W. Westerkamp, The Feynman integrand for the perturbed harmonic oscillator as a Hida distribution, *Ann. Phys.* **4** (1995) 53-67.

[61] D. de Falco and D. C. Khandekar, Applications of white noise calculus to the computation of Feynman integrals, *Stoch. Process. Appl.* **29** (1988) 257-266.

[62] M. de Faria, J. Potthoff, and L. Streit, The Feynman integrand as a Hida distribution, *J. Math. Phys.* **32** (1991) 2123-2127.

[63] M. de Faria, M. J. Oliveira, and L. Streit, Feynman integrals for non-smooth and rapidly growing potentials, *J. Math. Phys.* **46** (2005) Art. No. 063505.

[64] J. A. Deutsch, The cholinergic synapse and the site of memory, *Science* **174** (1971) 788-794.

[65] B. S. DeWitt, Dynamical theory in curved spaces. I. A review of the classical and quantum action principles, *Rev. Mod. Phys.* **29** (1957) 377.

[66] B. S. DeWitt, *Dynamical Theory of Groups and Fields* (Gordon and Breach, New York, 1965).

[67] B. S. DeWitt, Quantum field theory in curved spacetime, *Phys. Rep.* **19** (1975) 295.

[68] C. DeWitt-Morette, P. Cartier, and A. Folacci, eds., *Functional Integration. Basics and Applications* (Plenum Press, New York, 1997).

[69] C. DeWitt-Morette, Feynman's path integral: Definition without limiting procedure, *Commun. Math. Phys.* **28** (1972) 47-67.

[70] C. DeWitt-Morette, A. Maheshwari, and B. Nelson, Path integration in nonrelativistic quantum mechanics, *Phys. Rep. C* **50** (1979) 255-372.

[71] P. A. M. Dirac, The Lagrangian in quantum mechanics, *Phys. Zeitschr. Sowjetunion* **3** (1933) 64.

[72] P. A. M. Dirac, *The Principles of Quantum Mechanics* (Clarendon Press, Oxford, 1958).

[73] M. Doi and S. F. Edwards, *The Theory of Polymer Dynamics* (Oxford Univ. Press, New York, 1986).

[74] L. G. Dominguez, R. A. Wennberg, W. Gaetz, D. Cheyne, O. C. Snead III, and J. L. Velazquez, Enhanced Synchrony in Epileptiform Activity? Local versus Distant Phase Synchronization in Generalized Seizures, *J. Neurosci.* **25** (2005) 8077.

[75] J. L. Doob, *Stochastic Processes* (John Wiley & Sons, Canada, 1953).

[76] J. S. Dowker, Quantum mechanics and field theory on multiply connected and on homogeneous spaces, *J. Phys. A: Math. Gen.* **5** (1972) 936.

[77] S. F. Edwards, Statistical mechanics with topological constraints. I. *Proc. Phys. Soc.* **91** (1967) 513-519.

[78] A. Einstein, Über die von der molekularkinetischen Theorie der Wärme geforderte Bewegung von in ruhenden Flüssigkeiten suspendierten Teilchen, *Ann. Phys.* **17** (1905) 549-560.

[79] C. J. Eliezer and A. Gray, A note on the time-dependent harmonic oscillator, *SIAM J. Appl. Math.* **30** (1976) 463-468.

[80] R. Erban and S. J. Chapman, Reactive boundary conditions for stochastic simulations of reaction–diffusion processes, *Phys. Biol.* **4** (2007) 16-28.

[81] A. Erdelyi, W. Magnus, F. Oberhettinger, F. G. Tricomi, *Tables of Integral Transforms*, vol. 1 (McGraw-Hill, New York, 1954).

[82] E. Farhi and S. Gutmann, The functional integral on the half-line, *Int. J. Mod. Phys.* **5** (1990) 3029-3051.

[83] R. P. Feynman, Space-time approach to non-relativistic quantum mechanics, *Rev. Mod. Phys.* **20** (1948) 367.

[84] R. P. Feynman, Mathematical formulation of the quantum theory of electromagnetic interaction, *Phys. Rev.* **80** (1950) 440. (see Appendix A).

[85] R. P. Feynman, An operator calculus having applications in quantum electrodynamics, *Phys. Rev.* **84** (1951) 108. (See Appendix D).

[86] R. P. Feynman and A. R. Hibbs, *Quantum Mechanics and Path Integrals* (McGraw-Hill, New York, 1965).

[87] V. Fock, Proper time in classical and quantum mechanics, *Phys. Zeit. Sow. Un.* **12** (1937) 404.

[88] A. D. Fokker, Die mittlere energie rotierender elektrischer dipole im strahlungsfeld, *Ann. Phys.* **43** (1914) 810-820.

[89] K. Fraedrich, R. Morison, and L. M. Leslie, Improved tropical cyclone track predictions using error recycling. *Meteorol. Atmos. Phys.* **74** (2000) 51–56.

[90] C. W. Gardiner, *Handbook of Stochastic Methods*, 2nd ed., (Springer-Verlag, Berlin, 1985).

[91] C. C. Gerry and V. A. Singh, Feynman path-integral approach to the Aharonov-Bohm effect, *Phys. Rev. D* **20** (1979) 2550-2554.

[92] G. L. Gerstein and B. Mandelbrot, Random walk models for the spike activity of a single neuron, *Biophys. J.* **4** (1964) 41.

[93] R. Gibrat, *Les Inégalités Economiques* (Sirey, Paris, 1933).

[94] S. Gilman and S. W. Newman, *Manter and Gatz's Essentials of Clinical Neuroanatomy and Neurophysiology*, 10th ed., F. A. Davis (2002).

[95] H. Goldstein, *Classical Mechanics* (Addison-Wesley, Reading, 1980).

[96] E. S. R. Gopal, *Statistical Mechanics and Properties of Matter* (John Wiley & Sons, New York, 1974).

[97] J. Gore, Z. Bryant, M. Nöllmann, M. U. Le, N. R. Cozzarelli, and C. Bustamante, DNA overwinds when stretched, *Nature* **442** (2006) 836-839.

[98] I. S. Gradshteyn and I. M. Ryzhik, *Table of Integrals, Series and Products*, 5th ed. (Academic Press, San Diego, 1994).

[99] C. Grosche and F. Steiner, *Handbook of Feynman Path Integrals* (Springer, Berlin, 1998).

[100] M. Grothaus, D. C. Khandekar, J. L. da Silva, and L. Streit, The Feynman integral for time-dependent anharmonic oscillators, *J. Math. Phys.* **38** (1997) 3278.

[101] T. Hall and S. Jewson, Statistical modeling of North Atlantic tropical cyclone tracks, *Tellus* **59A** (2007) 486-498.

[102] T. Hida, *Stationary Stochastic Processes* (Princeton Univ. Press, Princeton, 1970), p. 95.

[103] T. Hida, *Analysis of Brownian Functionals*, Carleton Mathematical Lecture Notes **13** (1975).

[104] T. Hida, *Brownian Motion* (Springer-Verlag, New York, 1980).

[105] T. Hida, The role of exponential functions in the analysis of generalized Brownian functionals, *Theor. Probab. Appl.* **27** (1983) 609-613.

[106] T. Hida, H. H. Kuo, J. Potthoff, L. Streit, *White Noise. An Infinite Dimensional Calculus* (Kluwer, Dordrecht, 1993).

[107] R. Ho and A. Inomata, Exact path integral treatment of the hydrogen atom, *Phys. Rev. Lett.* **48** (1982) 231-234.

[108] J. J. Hopfield, Neural networks and physical systems with emergent selective computational abilities. *Proc. Natl. Acad. Sci. USA* **79** (1982) 2554.

[109] H. E. Hurst, Long-term storage of reservoirs: an experimental study, *Transactions of the American Society of Civil Engineers* **116** (1951) 770-799.

[110] H. E. Hurst, A suggested statistical model of some time series which occur in nature, *Nature* (London) **180** (1957) 494.

[111] L. Ingber, Statistical mechanics of neocortical interactions: Path-integral evolution of short-term memory, *Phys. Rev. E* **49** (1994) 4652-4664.

[112] A. Inomata and V. A. Singh, Path integrals and constraints: particle in a box, *Phys. Lett. A* **80** (1980) 105-108.

[113] C. Itzykson and J. B. Zuber, *Quantum Field Theory* (McGraw-Hill, New York, 1985).

[114] J.-H. Jeon, N. Leijnse, L. B. Oddershede, and R. Metzler, Anomalous diffusion and power-law relaxation of the time averaged mean squared displacement in worm-like micellar solutions, *New J. Phys.* **15** (2013) 045011.

[115] E. R. Kandel, J. H. Schwartz, and T. M. Jessell, *Principles of Neuroscience* (McGraw-Hill, New York, 2000).

[116] J. W. Kantelhardt, E. Koscielny-Bunde, D. Rybski, P. Braun, A. Bunde, and S. Havlin, Long-term persistence and multifractality of precipitation and river runoff records, *J. Geophys. Res.* **111**: D01106 (2006) DOI: 10.1029/2005JD005881.

[117] T. Karagiannis, M. Molle, and M. Faloutsos, Long-range dependence-ten years of internet traffic modeling, *IEEE Internet Computing* **8** (2004) 57-64.

[118] M. A. Kayed and A. Inomata, Exact path-integral solution of the Dirac-Coulomb problem, *Phys. Rev. Lett.* **53** (1984) 107-110.

[119] J. B. Keller and D. W. McLaughlin, The Feynman integral, *Am. Math. Monthly* **82** (1975) 451-465.

[120] J. A. Kelso, *Dynamic Patterns: the Self-Organization of Brain and Behavior* (MIT Press, Cambridge, 1995).

[121] D. C. Khandekar and S. V. Lawande, Exact propagator for a time-dependent harmonic oscillator with and without a singular potential, *J. Math. Phys.* **16** (1975) 384-388.

[122] D. C. Khandekar and L. Streit, Constructing the Feynman integrand, *Ann. Phys.* (Leipzig) **1** (1992) 46-55.

[123] J. Klafter, S. C. Lim, R. Metzler, eds., *Fractional Dynamics* (World Scientific, Singapore, 2012).

[124] R. Klages, G. Radons, and I. M. Sokolov, *Anomalous Transport* (Wiley-VCH, Weinheim, 2008).

[125] J. R. Klauder and I. Daubechies, Measures for path integrals, *Phys. Rev. Lett.* **48** (1982) 117-120.

[126] J. R. Klauder and I. Daubechies, Quantum mechanical path integrals with Wiener measures for all polynomial Hamiltonians, *Phys. Rev. Lett.* **52** (1984) 1161-1164.

[127] J. R. Klauder and B.-S. Skagerstam, *Coherent States. Applications in Physics and Mathematical Physics* (World Scientific, Singapore, 1985).

[128] J. A. Kleim *et al.*, Synapse formation is associated with memory storage in the cerebellum, *Proc. Natl. Acad. Sci.* **99** (2002) 13228-13231.

[129] H. Kleinert, *Path Integrals in Quantum Mechanics, Statistics, Polymer Physics, and Financial Markets* (World Scientific, Singapore, 2009).

[130] H. Koizumi, Y. Yamashita, A. Maki, T. Yamamoto, Y. Ito, H. Itagaki, and

R. Kennan, Higher-order brain function analysis by trans-cranial dynamic near infrared spectroscopy imaging, *J. Biomed. Opt.* **4** (1999) 403-413.

[131]   A. Kumar, S. Rotter, and A. Aertsen, Spiking activity propagation in neuronal networks: reconciling different perspectives on neural coding, *Nature Rev. Neurosci.* **11** (2010) 615.

[132]   T. Kuna, L. Streit, and W. Westerkamp, Feynman integrals for a class of exponentially growing potentials, *J. Math. Phys.* **39** (1998) 4476-4491.

[133]   H. H. Kuo, *White Noise Distribution Theory* (CRC, Boca Raton, FL, 1996).

[134]   H. H. Kuo, On Laplacian operators of generalized Brownian functionals, in *Stochastic Processes and their Applications,* LNM 1203, eds. K. Ito and T. Hida (Springer-Verlag, Berlin, 1986), pp. 119-128.

[135]   H. H. Kuo, Lectures on white noise analysis, *Soochow J. Math.* **18** (1992) 229-300.

[136]   F. A. Labra, P. A. Marquet, and F. Bozinovic, Scaling metabolic rate fluctuations, *Proc. Natl. Acad. Sci.* **104** (2007) 10900-10903.

[137]   M. G. Laidlaw and C. M. DeWitt, Feynman functional integrals for systems of indistinguishable particles, *Phys. Rev. D* **3** (1971) 1375.

[138]   C. Laing and G. J. Lord, eds., *Stochastic Methods in Neuroscience* (Oxford Univ. Press, 2009).

[139]   A. Lascheck, P. Leukert, L. Streit, and W. Westerkamp, Quantum mechanical propagators in terms of Hida distributions, *Rep. Math. Phys.* **33** (1993) 221-232.

[140]   A. Lascheck, P. Leukert, L. Streit, and W. Westerkamp, More about Donsker's delta function, *Soochow J. Math.* **20** (1994) 401-418.

[141]   J. Lebovits and J. L. Véhel, White noise-based stochastic calculus with respect to multifractional Brownian motion, *Stoch.: Int. J. Probab. Stoch. Proc.* **86** (2014) 87-124.

[142]   P. Lévy, *Le Mouvement Brownien* (Gauthier-Villars, Paris, 1954).

[143]   M. Li and S. C. Lim, Modeling network traffic using generalized Cauchy process, *Physica A: Statist. Mech. Appl.* **387** (2008) 2584-2594.

[144]   S. C. Lim and S. V. Muniandy, Self-similar Gaussian processes for modeling anomalous diffusion, *Phys. Rev. E* **66** (2002) 021114.

[145]   S. C. Lim and V. M. Sithi, Asymptotic properties of the fractional Brownian motion of Riemann-Liouville type, *Phys. Lett. A* **206** (1995) 311-317.

[146]   T. Lionnet, S. Joubaud, R. Lavery, D. Bensimon, and V. Croquette, Wringing out DNA, *Phys. Rev. Lett.* **96** (2006) 178102.

[147]   R. Mahnke, J. Kaupužs, and I. Lubashevsky, *Physics of Stochastic Processes* (Wiley VCH, Weinheim, 2009)

[148]   B. Mandelbrot and J. W. van Ness, Fractional Brownian motions, fractional noises and applications, *SIAM Rev.* **10**(4) (1968) 422-437.

[149]   S. J. Martin, P. D. Grimwood, and R. G. Morris, Synaptic plasticity and memory: An evaluation of the hypothesis, *Annu. Rev. Neurosci.* **23** (2000) 649-711.

[150]   T. G. Mason, Estimating the viscoelastic moduli of complex fluids using the generalized Stokes-Einstein equation, *Rheol. Acta* **39** (2000) 371-378.

[151]   T. G. Mason and D. A. Weitz, Optical measurements of frequency-

dependent linear viscoelastic moduli of complex fluids, *Phys. Rev. Lett.* **74** (1995) 1250-1253.

[152] D. W. McLaughlin and L. S. Schulman, Path integrals in curved spaces, *J. Math. Phys.* **12** (1971) 2520.

[153] R. Metzler and J. Klafter, The random walks's guide to anomalous diffusion, *Phys. Rep.* **339** (2000) 1.

[154] T. Meuel, G. Prado, F. Seychelles, M. Bessafi, and H. Kellay, Hurricane track forecast cones from fluctuations, *Sci. Rep.* **2** (2012) 446; DOI:10.1038/srep00446.

[155] K. S. Miller and B. Ross, *An Introduction to the Fractional Calculus and Fractional Differential Equations* (John Wiley & Sons, New Jersey, 1993).

[156] Y. Mishura, *Stochastic Calculus for a Fractional Brownian Motion and Related Processes* (Springer-Verlag, Berlin, 2008).

[157] C. W. Misner, Feynman quantization of general relativity, *Rev. Mod. Phys.* **29** (1957) 497.

[158] M. S. Mohaved, G. R. Jafari, F. Ghasemi, S. Rahvar, M. R. R. Tabar, Multifractal detrended fluctuation analysis of sunspot time series, *J. Stat. Mech.: Theory Exp.* **2006** (2006) P02003. DOI: 10.1088/1742-5468/2006/02/P02003.

[159] C. Morette, On the definition and approximation of Feynman's path integrals, *Phys. Rev.* **81** (1951) 848.

[160] S. Nadkarni and P. Jung, Modeling synaptic transmission of the tripartite synapse, *Phys. Biol.* **4** (2007) 1-9.

[161] Y. Nambu, The use of the proper time in quantum electrodynamics I, *Prog. Theor. Phys.* **V** (1950) 82-94.

[162] T. I. Netoff and S. J. Schiff, Decreased neuronal synchronization during experimental seizures, *J. Neurosci.* **22** (2002) 7297-7307.

[163] N. Obata, *White Noise Calculus and Fock Space*, Lecture Notes in Mathematics, Vol. 1577 (Springer, Berlin, 1994).

[164] OECD-CERI, Understanding the Brain: Towards a New Learning Science (Organization for Economic Cooperation and Development, 2002).

[165] I. E. Ohiorhenuan, F. Mechler, K. P. Purpura, A. M. Schmid, Q. Hu, and J. D. Victor, Sparse coding and high-order correlations in fine-scale cortical networks, *Nature* **466** (2010) 617-621.

[166] T. Ohira and J. D. Cowan, Path integrals for stochastic neurodynamics, in *Proc. World Congress on Neural Networks*, San Diego, June 1994 (Sony Computer Science Lab. Inc., 1995).

[167] G. J. Papadopoulos and J. T. Devreese, Path integral solutions of the Dirac equation, *Phys. Rev. D* **13** (1976) 2227.

[168] G. Parisi and Y. S. Wu, Perturbation theory without gauge fixing, *Scientia Sinica* **24** (1981) 483-496.

[169] C.-K. Peng, S. V. Buldyrev, A. L. Goldberger, S. Havlin, F. Sciortino, M. Simons, and H. E. Stanley, Long-range correlations in nucleotide sequences, *Nature* **356** (1992) 168-170.

[170] M. Peshkin and A. Tonomura, *The Aharonov-Bohm Effect* (Springer, Berlin, 1989).

[171] A. M. Petersen, J. Tenenbaum, S. Havlin, and H. E. Stanley, Statistical laws governing fluctuations in word use from word birth to word death, *Sci. Rep.* **2** (2012) 313-321.

[172] A. M. Petersen, M. Riccaboni, H. E. Stanley, and F. Pammolli, Persistence and uncertainty in the academic career, *Proc. Natl. Acad. Sci.* **109** (2012) 5213-5218.

[173] L. Pietronero, Survival probability for kinetic self-avoiding walks, *Phys. Rev. Lett.* **55** (1985) 2025-2027.

[174] M. Planck, Über einen satz der statistischen dynamik und seine erweiterung in der quantentheorie, *Sitzber. Preuss. Akad. Wiss.* (1917) 324.

[175] J. Potthoff and L. Streit, A characterization of Hida distributions, *J. Funct. Anal.* **101** (1991) 212-229.

[176] H. Qian, Fractional Brownian motion and fractional Gaussian noise, in *Processes with Long-range Correlations: Theory and Applications, Lecture Notes in Physics*, Vol. 621, eds. G. Rangarajan and M. Z. Ding (Springer, Berlin, 2003), pp. 22-33.

[177] B. Qian and K. Rasheed, Hurst exponent and financial market predictability, *Proceedings of the 2nd IASTED International Conference on Financial Engineering and Applications* (Acta Press, Calgary, 2004), pp. 203-209.

[178] P. Ramond, *Field Theory - A Modern Primer* (Benjamin/Cummings Publ., Reading, 1981).

[179] R. C. Ramos, Jr., C. C. Bernido, and M. V. Carpio-Bernido, Path-integral treatment of ring-shaped topological defects, *J. Phys. A: Math Gen.* **27** (1994) 8251.

[180] J. C. Reboredo, M. A. Rivera-Castro, J. G. V. Miranda, and R. Garcia-Rubio, How fast do stock prices adjust to market efficiency? Evidence from a detrended fluctuation analysis, *Physica A* **392** (2013) 1631-1637.

[181] H. Risken, *The Fokker-Planck Equation* (Springer-Verlag, Berlin, 1996).

[182] J. B. Salig, Jr., M. V. Carpio-Bernido, C. C. Bernido, and J. B. Bornales, On neuron membrane potential distributions for voltage and time dependent current modulation, *Int. J. Mod. Phys. Conf. Ser.* **17** (2012) 19-22.

[183] M. Sanhueza, G. Fernandez-Villalobos, I. S. Stein, G. Kasumova, P. Zhang, K. U. Bayer, N. Otmakhov, J. W. Hell, and J. Lisman, Role of the CaMKII/NMDA receptor complex in the maintenance of synaptic strength, *J. Neurosci.* **31** (2011) 9170.

[184] Y. C. Sasaki, Y. Okumura, S. Adachi, H. Suda, Y. Taniguchi, and N. Yagi, Picometer-scale dynamical x-ray imaging of single DNA molecules, *Phys. Rev. Lett.* **87** (2001) 248102.

[185] L. S. Schulman, *Techniques and Applications of Path Integration* (Wiley, New York, 1981).

[186] J. Schwinger, On gauge invariance and vacuum polarization, *Phys. Rev.* **82** (1951) 664.

[187] R. E. Shaw, E. E. Kadar, and M. T. Turvey, The job description of the cerebellum and a candidate model of its "tidal wave" function, *Behavioral and Brain Sciences* **20** (1997) 265.

[188] V. M. Sithi and S. C. Lim, On the spectra of Riemann-Liouville fractional Brownian motion, *J. Phys. A: Math. Gen.* **28** (1995) 2995-3003.

[189] D. E. Smith, T. T. Perkins, and S. Chu, Self-diffusion of an entangled DNA molecule by reptation, *Phys. Rev. Lett.* **75** (1995) 4146.

[190] M. A. Smith and A. Kohn, Spatial and temporal scales of neuronal correlation in primary visual cortex, *J. Neurosci.* **28** (2008) 12591-12603.

[191] T. Sobayo, A. Fine, E. Gunnar, C. Kazlauskas, D. Nicholls, and D. Mogul, Synchrony dynamics across brain structures in limbic epilepsy vary between initiation and termination phases of seizures, *IEEE Trans Biomed. Eng.* (2012). DOI:10.1109/TBME.2012.2189113.

[192] I. M. Sokolov, J. Klafter, and A. Blumen, Fractional kinetics, *Physics Today*, November (2002) 48-54.

[193] R. J. Staba, C. L. Wilson, A. Bragin, I. Fried, and J. Engel, Jr., Sleep states differentiate single neuron activity recorded from human epileptic hippocampus, entorhinal cortex, and subiculum, *J. Neurosci.* **22** (2002) 5694-5704.

[194] M. H. R. Stanley, L. A. N. Amaral, S. V. Buldyrev, S. Havlin, H. Leschhorn, P. Maass, M. A. Salinger, and H. E. Stanley, Scaling behavior in the growth of companies, *Nature* **379** (1996) 804-806.

[195] L. Streit, Feynman paths, sticky walls, white noise, in *A Garden of Quanta, Essays in Honor of Hiroshi Ezawa*, eds. J. Arafune et al. (World Scientific, Singapore, 2003), pp. 105-113.

[196] L. Streit and T. Hida, Generalized Brownian functionals and the Feynman integral, *Stoch. Proc. Appl.* **16** (1983) 55-69.

[197] L. Streit, The Feynman integral - answers and questions, in *Proc. of the 1st Jagna International Workshop on Advances in Theoretical Physics,* eds. C. C. Bernido and M. V. Carpio-Bernido (Central Visayan Institute, Jagna, 1996), pp. 188-199.

[198] L. Stryer, *Biochemistry*, 4th ed. (W. H. Freeman and Company, New York, 1998).

[199] J. Sutton, Gibrat's legacy, *J. Econ. Lit.* **35** (1997) 40-59.

[200] M. Tassieri, R. M. L. Evans, R. L. Warren, N. J. Bailey, and J. M. Cooper, Microrheology with optical tweezers: data analysis, *New J. Phys.* **14** (2012) 115032.

[201] S. F. Tolić-Norrelykke, M. B. Rasmussen, F. S. Pavone, K. Berg-Sorensen, and L. B. Oddershede, Stepwise bending of DNA by a single TATA-box binding protein, *Biophys. J.* **90** (2006) 3694.

[202] T. Trappenberg, *Fundamentals of Computational Neuroscience*, 2nd ed. (Oxford Univ. Press, Oxford, 2010).

[203] V. Uchaikin and R. Sibatov, *Fractional Kinetics in Solids* (World Scientific, Singapore, 2013).

[204] G. E. Uhlenbeck and L. S. Ornstein, On the theory of Brownian motion, *Phys. Rev.* **36** (1930) 823-841.

[205] D. Voet, J. G. Voet, and C. W. Pratt, *Fundamentals of Biochemistry*, 2nd ed. (John Wiley and Sons, 2006), p. 182.

[206] N. Voges, A. Schüz, A. Aertsen, and S. Rotter, A modeler's view on the

spatial structure of intrinsic horizontal connectivity in the neocortex, *Progr. Neurobiol.* **92** (2010) 277-292.

[207] K. G. Wang and C. W. Lung, Long-time correlation effects and fractal Brownian motion, *Phys. Lett. A* **151** (1990) 119-121.

[208] M. C. Wang and G. E. Uhlenbeck, On the theory of Brownian motion II, *Rev. Mod. Phys.* **17** (1945) 323-342.

[209] G. Werner, Fractals in the nervous system: Conceptual implications for theoretical neuroscience, *Front. Physiol.* **1** (2010) 1-28, DOI: 10.3389/fphys.2010.00015.

[210] H. S. Wio, *Path Integrals for Stochastic Processes - An Introduction* (World Scientific, Singapore, 2013).

[211] E. Wong, *Introduction to Random Processes* (Springer Verlag, Penn., 1983).

[212] W. A. Woodward, H. L. Gray, and A. C. Elliot, *Applied Time Series Analysis* (CRC Press, Boca Raton, 2011).

[213] J. S. Yang, S. Wallin, and E. I. Shakhnovich, Universality and diversity of folding mechanics for three-helix bundle proteins, *Proc. Nat'l. Acad. Sci. (USA)* **105** (2008) 895-900.

[214] K. Yasue, M. Jibu, T. Misawa, and J.-C. Zambrini, Stochastic neurodynamics, *Ann. Inst. Statist. Math.* **40** (1988) 41-59.

[215] T. Yoshida, Universal dependence of the mean square displacement in equilibrium point vortex systems without boundary conditions, *J. Phys. Soc. Jpn.* **78** (2009) 024004.

[216] B. Young and J. W. Heath, *Wheater's Functional Histology*, 4th ed. (Churchill Livingston, Harcourt Pub. Lim., London, 2000).

[217] W. Yu, K. Chung, M. Cheon, M. Heo, K.-H. Han, S. Ham, and I. Chang, Cooperative folding kinetics of BBL protein and peripheral subunit-binding domain homologues, *Proc. Nat'l. Acad. Sci. (USA)* **105** (2008) 2397-2402.

Index

Printed in the United States
By Bookmasters